PERFORMANCE EVALUATION OF MICRO IRRIGATION MANAGEMENT

Principles and Practices

Innovations and Challenges in Micro Irrigation

VOLUME 3

PERFORMANCE EVALUATION OF MICRO IRRIGATION MANAGEMENT

Principles and Practices

Edited by

Megh R. Goyal, PhD, PE, Senior Editor-in-Chief

Apple Academic Press Inc. | Apple Academic Press Inc.
3333 Mistwell Crescent | 9 Spinnaker Way
Oakville, ON L6L 0A2 | Waretown, NJ 08758
Canada | USA

©2017 by Apple Academic Press, Inc.

First issued in paperback 2021

Exclusive worldwide distribution by CRC Press, a member of Taylor & Francis Group
No claim to original U.S. Government works

ISBN 13: 978-1-77463-601-5 (pbk)
ISBN 13: 978-1-77188-320-7 (hbk)

Library and Archives Canada Cataloguing in Publication

Performance evaluation of micro irrigation management : principles and practices / edited by Megh R. Goyal, PhD, PE, senior editor-in-chief.

(Innovations and challenges in micro irrigation ; volume 3)
Includes bibliographical references and index.
Issued in print and electronic formats.
ISBN 978-1-77188-320-7 (hardcover).--ISBN 978-1-77188-321-4 (pdf)
1. Microirrigation--Evaluation. 2. Microirrigation--Management. 3. Water in agriculture.
I. Goyal, Megh Raj, author, editor II. Series: Innovations and challenges in micro irrigation; v. 3
S619.T74P47 2016   631.5'87   C2016-904011-9   C2016-904012-7

Library of Congress Cataloging-in-Publication Data

Names: Goyal, Megh Raj, editor.
Title: Performance evaluation of micro irrigation management : principles and practices / editor: Megh R. Goyal.
Other titles: Innovations and challenges in micro irrigation ; [v. 3]
Description: Waretown, NJ : Apple Academic Press, 2017. | Series: Innovations and chal-lenges micro irrigation ; [volume 3] | Includes bibliographical references and index.
Identifiers: LCCN 2016026202 (print) | LCCN 2016027151 (ebook) | ISBN 9781771883207 (hardcover : alk. paper) | ISBN 9781771883214 ()
Subjects: LCSH: Microirrigation--Research. | Microirrigation--Evaluation.
Classification: LCC S619.T74 P42 2017 (print) | LCC S619.T74 (ebook) | DDC 631.5/87--dc23
LC record available at https://lccn.loc.gov/2016026202

Apple Academic Press also publishes its books in a variety of electronic formats. Some content that ap-pears in print may not be available in electronic format. For information about Apple Academic Press products, visit our website at **www.appleacademicpress.com** and the CRC Press website at **www.crc-press.com**

CONTENTS

LIST OF CONTRIBUTORS

R. B. Biniwale, PhD
Principal Scientist, National Environmental Engineering Research Institute, Nagpur, India

Vishal K. Chavan, PhD
Assistant Professor and Senior Research Fellow in SWE, AICRP for Dryland Agriculture, Dr. PDKV, Akola, Maharashtra, India. Mobile: +91-7875689897, E-mail: vchavan2@gmail.com; Website: www.pdkv.ac.in

M. M. Deshmukh, PhD
Professor, Department of Irrigation and Drainage Engineering, Dr. Panjabrao Deshmukh Krishi Vidyapeeth, Krishinagar PO, Akola (MS) 444104, India

S. M. Ghawade, PhD
Department of Irrigation and Drainage Engineering, Dr. Panjabrao Deshmukh Krishi Vidyapeeth, Akola 444104, India

Megh R. Goyal, PhD, PE
Retired Professor in Agricultural and Biomedical Engineering, University of Puerto Rico – Mayaguez Campus; and Senior Technical Editor-in-Chief in Agriculture Sciences and Biomedical Engineering, Apple Academic Press Inc., PO Box 86, Rincon – PR – 00677, USA. E-mail: goyalmegh@gmail.com

Eric W. Harmsen, PhD
Professor, Department of Agricultural and Biosystems Engineering, University of Puerto Rico, Mayaguez Campus, Mayaguez, Puerto Rico 00681 USA. Phone: +1-787-955-5102; E-mail: eric.harmsen@upr.edu, harmsen1000@gmail.com; Website: www.pragwater.com;

Vishal Keshaorao Ingle, PhD Scholar (SWCE)
Division of Agricultural Engineering, Indian Agricultural Research Institute, New Delhi – 110012; Assistant Professor (Soil and water conservation Engineering), College of Agriculture, Aland Road, Gulbarga-02, Karnataka, India. E-mail: vishal4644@gmail.com

M. U. Kale, PhD
Professor, Department of Irrigation and Drainage Engineering, Dr. Panjabrao Deshmukh Krishi Vidyapeeth, Akola 444104 (MS), India. E-mail: kale921@gmail.com

Richard Koech, PhD
School of Environmental and Rural Science, University of New England, Armidale NSW 2351, Australia. Phone: +61–267735221; E-mail: rkoech@une.edu.au or richardkoech@hotmail.com.

J. N. Lokhande, PhD
Professor, Department of Irrigation and Drainage Engineering, Dr. Panjabrao Deshmukh Krishi Vidyapeeth, Akola 444104 (MS), India.

John Mecikalski, PhD
Associate Professor, Department of Atmospheric Sciences, University of Alabama in Huntsville, 320 Sparkman Drive, NSSTC, Huntsville, AL 35806, USA. E-mail: johnm@nsstc.uah.edu

Ashok R. Mhaske, PhD
Associate Professor, Soil Water Conservation Eng. Agric. College, Dr. PDKV, Nagpur, India. E-mail: mhaskear@gmail.com

A. M. Michael, PhD
Former Professor and Project Director, Water Technology Center, IARI, New Delhi, Director, IARI, New Delhi; and Ex-Vice-Chancellor, Kerala Agricultural University, Trichur, Kerala; Present address: Dr. A. M. Michael, 34/81, V.P. Marikar Road, Edappally North P.O., Kochi–682024, Kerala, India

A. K. Mishra, PhD
Division of Agricultural Engineering, Indian Agricultural Research Institute, New Delhi 110012, India

Muñoz-Muñoz, Miguel A., PhD
Ex-President of University of Puerto Rico, University of Puerto Rico, Mayaguez Campus, College of Agriculture Sciences, Call Box 9000, Mayaguez, PR 00681-9000, Tel.: 787-265-3871, E-mail: miguel.munoz3@upr.edu

Ian Nuberg, PhD
School of Agriculture, Food and Wine, University of Adelaide, Frome Street, 5005, Adelaide, SA 5005, Australia. Phone: +61-883130527; E-mail: ian.nuberg@adelaide.edu.au

D. B. Palwe, PhD
Professor, Department of Irrigation and Drainage Engineering, Dr. Panjabrao Deshmukh Krishi Vidyapeeth, Akola 444104 (MS), India

David Pezzaniti, PhD
Australian Irrigation and Hydraulic Technology Centre, University of South Australia, PO Box 2471, Adelaide, Australia. Phone: +61-8-830-23652; E-mail: David.Pezzaniti@unisa.edu.au

C. Prabakaran, PhD
Assistant Professor (Environmental Sciences), Agricultural Research Station, Tamil Nadu Agricultural University (TNAU), Thirupathisaram–629901 E-mail: prabaka_123@yahoo.co.in, prabakarancinnusamy@rediff.com

Abdelraouf Ramadan Eid Abdelghan, PhD
Associate Professor of Agricultural Engineering, Water Relations and Field Irrigation Department, Agricultural and Biological Research Division, National Research Centre – Dokki – Giza, Egypt, 33 El-Buhouth St., Dokki, Cairo, Egypt, P.O. 12311, Mobile: (0202) 01146166904, E-mail: abdelrouf2000@yahoo.com

Victor J. Reventos
Undergraduate Research Assistant, Department of Computer and Electrical Engineering, Puerto Rico, Mayaguez, PR 00681. E-mail: victor.reventos@upr.edu

A. S. Talokar, PhD
Department of Irrigation and Drainage Engineering, Dr. Panjabrao Deshmukh Krishi Vidyapeeth, Akola 444104 (MS), India

S. B. Wadatkar, PhD
Department of Irrigation and Drainage Engineering, Dr. Panjabrao Deshmukh Krishi Vidyapeeth, Akola 444104 (MS), India

Isa Yunusa, PhD
School of Environmental and Rural Science, University of New England, Armidale NSW 2351, Australia. Phone: +61-2 6773-2436; E-mail: isa.yunusa@une.edu.au

LIST OF ABBREVIATIONS

Θ_w	dry weight basis
°C	degree Celsius
AE	application ffficiency
ASABE	American Society of Agricultural and Biological Engineers
BD	bulk density
CDI	continuous drip irrigation
CRE	clogging ratio of emitters
CU	coefficient of uniformity
DIS	drip irrigation system
DU	distribution uniformity
EPAN	pan evaporation
Es	bare soil evaporation
ET	evapotranspiration
et al.	and others
ET_c	crop evapotranspiration
etc.	etcetera
EU	emission uniformity
EUE	cnergy use efficiency
EUE_{potato}	energy use efficiency of potato
FAO	Food and Agricultural Organization
FC	field capacity
Fed	feddan = 0.42 ha = 4200 m^2
FUE	fertilizer use efficiency
gpm	gallons per minute
h day^{-1}	hours per day
hrs	hours
i.e.	that is
ICAR	Indian Council of Agriculture Research
IR	water intake rate into the soil

ISAE	Indian Society of Agricultural Engineers
k_c	crop coefficient
Kg	kilograms
km h^{-1}	kilometer per hour
K_p	pan coefficient
kPa	kilopascal
Ks	hydraulic conductivity
KUE_{potato}	potassium use efficiency of potato
LAI	leaf area index
lps	liters per second
m s^{-1}	meter per second
MAE	mean absolute error
MC	moisture content
MCRZ	moisture content in root zone
min	minute
mm day^{-1}	millimeter per day
MSL	mean sea level
NI	net income
NUE_{potato}	nitrogen use efficiency of potato
PDI	pulse drip irrigation
PET	potential evapotranspiration
pH	acidity/alkalinity measurement scale
PM	Penman-Monteith
PSDI	pulse surface drip irrigation
PSSDI	pulse subsurface drip irrigation
PUE_{potato}	phosphate use efficiency of potato
PWP	permanent wilting point
R^2	coefficient of determination
RA	extraterrestrial radiation
RF	rainfall
RH	relative humidity
RMSE	root mean square error
RS	solar radiation
SDI	surface drip irrigation
SR	solar radiation
SSDI	subsurface drip irrigation

SSH	sunshine hours
SW	saline water
TDMC	tuber dry matter content
TE	transpiration efficiency
TMAX	maximum temperature
TMIN	minimum temperature
TR	temperature range
TSD	tuber specific density
TTC	tuber total carbohydrates
TUE	transpiration use efficiency
USDA	US Department of Agriculture
USDA-SCS	US Department of Agriculture-Soil Conservation Service
viz.	namely
WS	wind speed
$WSV_{\geq 100\%FC}$	wetted soil volume (more than or equal 100% of field capacity)
WUE	water use efficiency
WUE_{potato}	water use efficiency for potato
YP	yield of potato

LIST OF SYMBOLS

A	cross sectional flow area (L^2)
AW	available water ($\Theta_w\%$)
Cp	specific heat capacity of air, in J/(g·°C)
CV	coefficient of variation
Δ	slope of the vapor pressure curve (kPa°C^{-1})
e	vapor pressure, in kPa
e_a	actual vapor pressure (kPa)
E	evapotranspiration rate, in g/(m²σ)
Ecp	cumulative class-A pan evaporation for two consecutive days (mm)
E_i	irrigation efficiency of drip system
E_p	Pan evaporation as measured by Class-A pan evaporimeter (mm/day)
Es	saturation vapor pressure, in kPa
E_{pan}	Class A pan evaporation
e_s	saturation vapor pressure (kPa)
$e_s - e_a$	vapor pressure deficit (kPa)
ET	evapotranspiration rate, in mm/year
ETa	reference ET, in the same water evaporation units as Ra
ETc	crop-evapotranspiration (mm/day)
ET_o	the reference evapotranspiration obtained using the Penman-Monteith method, (mm/day)
ET_{pan}	the pan evaporation-derived evapotranspiration
EU	emission uniformity
F	flow rate of the system (gpm)
F.C.	field capacity (v/v, %)
G	soil heat flux at land surface, in W/m²
H	the plant canopy height in meter
h	the soil water pressure head (L)
I	the infiltration rate at time t (mm/min)
K	the unsaturated hydraulic conductivity (LT^{-1})

K_c	crop coefficient
Kp	Pan factor
K_p	Pan coefficient
n	number of emitters
Pa	atmospheric pressure, in Pa
Q	flow rate in gallons per minute
q	the mean emitter discharges of each lateral (lh^{-1})
	the mean emitter discharge for all of the emitters
q_{lq}	the average emitter discharge from the lower quartile of sampled emitters
Q_{avg}	overall average of emitter discharge (l/h)
q_{ini}	mean discharge of new emitters at the same operating pressure of 98.06 kPa (lh^{-1})
Q_q	average low-quarter emitter discharge (l/h)
r_a	aerodynamic resistance (s m^{-1})
Ra	extraterrestrial radiation, in the same water evaporation units as ETa
R_e	effective rainfall depth (mm)
R_n	net radiation at the crop surface (MJ m^{-2}day^{-1})
Rs	incoming solar radiation on land surface, in the same water evaporation units as ETa
RO	surface runoff
r_s	the bulk surface resistance (s m^{-1})
V	volume of water required (liter/day/plant)
V_{id}	irrigation volume applied in each irrigation (liter tree^{-1})
W_p	fractional wetted area
γ	psychrometric constant (kPa°C^{-1})
θ	volumetric soil water content (L^{3}L^{-3})
$\theta(h)$	the soil water retention (L^{3}L^{-3}),
$\theta\rho$	the residual water content (L^{3}L^{-3})
θ_s	the saturated water content (L^{3}L^{-3})
θ_v	the vertically averaged volumetric soil moisture content
θ_{vol}	a volumetric moisture content (cm^3/cm^3)
λ	latent heat of vaporization (MJ κγ^{-1})
λE	latent heat flux, in W/mo
ρ_a	mean air density at constant pressure (kg m^{-3})
ρ_b	the bulk density (gm/cm^3)

PREFACE

Due to increased agricultural production, irrigated land has increased in the arid and sub-humid zones around the world. Agriculture has started to compete for water use with industries, municipalities and other sectors. This increasing demand along with increments in water and energy costs have made it necessary to develop new and innovative technologies for the adequate management of water. The intelligent use of water for crops requires understanding of evapotranspiration processes and use of efficient irrigation methods.

Micro irrigation is sustainable and is one of the best management practices. I attended the 17th Punjab Science Congress on February 14–16, 2014 at Punjab Technical University in Jalandhar and the 49th Annual Convention of Indian Society of Agricultural Engineers (ISAE) on February 22–25, 2015 at Punjab Agricultural University in Ludhiana. At these conventions, I was shocked to learn that the underground water table has lowered to a critical level in Punjab. My father-in-law, Mahasha Partigya Pal (he does not hold a university degree and in not involved in farming) in Dhuri, told me that his family bought the 0.10 acres of land in the city for US$100.00 in 1942 because the water table was at 2 feet depth. In 2012, it was sold for US$233,800 because the water table had dropped to greater than 100 feet. This has been due to luxury use of water by wheat-paddy farmers. This implies that even a layperson is able to identify the problems and benefits of water scarcity. The water crisis is similar in other countries, including Puerto Rico where I live. Year 2015 has been a drought year in most of the countries of the world. Who should be blamed for water scarcity: God or human beings? What has caused drought? I leave it to the reader to answer. We can, however, conclude that the problem of water scarcity is rampant globally, creating the urgent need for water conservation. The use of micro irrigation systems is expected to result in water savings, and increased crop yields in terms of volume and quality.

Our planet will not have enough potable water for a population of >10 billion persons in 2115. The situation will be further complicated by multiple factors that will be adversely affected by global warming. The website at http://www.un.org/waterforlifedecade/scarcity.shtml indicates, *"Water scarcity already affects all continents. Around 1.2 billion people, or almost one-fifth of the world's population, live in areas of physical scarcity, and 500 million people are approaching this situation. Another 1.6 billion people, or almost one quarter of the world's population, face economic water shortage (where countries lack the necessary infrastructure to take water from rivers and aquifers). Water scarcity is among the main problems to be faced by many societies and the World in the 21st century. Water use has been growing at more than twice the rate of population increase in the last century, and, although there is no global water scarcity as such, an increasing number of regions are chronically short of water. Water scarcity is both a natural and a human-made phenomenon. There is enough freshwater on the planet for seven billion people but it is distributed unevenly and too much of it is wasted, polluted and unsustainably managed."*

Micro/drip/trickle/pulse irrigation system can partially help to alleviate this rampant crisis, because of high irrigation efficiency. Every day, news on the importance of micro irrigation appear around the world indicating that government agencies at central/state/local levels, research and educational institutions, industry, sellers and others are aware of the urgent need to adopt micro irrigation technology that can have an irrigation efficiency up to 90% compared to 30–40% for the conventional irrigation systems.

It is important to adopt a suitable drip irrigation system to grow agricultural crops, space plants, forest trees, landscape plants and shrubs, and garden plants because all vegetation requires different water intake. For better results, one should plan and install a proper irrigation system for the land under consideration. Micro irrigation is one of the most efficient watering methods, as it can save water and give better quality of products. The trickle irrigation system can be designed and adapted to varying irrigation needs for: the arid, semi-arid and humid regions; wide range of crops; climatic and soil conditions. Drip irrigation can save our planet from the water scarcity.

The trickle irrigation design must be carried out by a professional registered engineer who is qualified and has the necessary knowledge. It is not a job of a layperson. Investment in the design phase will pay off in the long run. In November of 1979, a hydraulic technician came to my office and tried to convince me that he could design drip irrigation better than the engineer. One of his systems at a 500-hectare vegetable farm in Puerto Rico failed during the first crop. I helped to save this farm from total failure. We had to do the necessary modifications to the existing design and replace the necessary parts. I recommend 100% to consult an engineer to design the drip irrigation system.

Micro irrigation, also known as trickle irrigation or drip irrigation or localized irrigation or high frequency or pressurized irrigation, is an irrigation method that saves water and fertilizer by allowing water to drip slowly to the roots of plants, either onto the soil surface or directly onto the root zone, through a network of valves, pipes, tubing, and emitters. It is done through narrow tubes that deliver water directly to the base of the plant. It is a system of crop irrigation involving the controlled delivery of water directly to individual plants and can be installed on the soil surface or subsurface.

The other important benefits of using micro irrigation systems include expansion in the area under irrigation, water conservation, optimum use of fertilizers and chemicals through water, and decreased labor costs, among others. Micro irrigation systems are often used in farms and large gardens, but are equally effective in the home garden or even for houseplants or lawns. They are easily customizable and can be set up even by inexperienced gardeners. Putting a drip system into the garden is a great do-it-yourself project that will ultimately save the time and help the plants grow. It is equally used in landscaping and in green cities.

The mission of this book volume is to serve as a reference manual for graduate and undergraduate students of agricultural, biological and civil engineering as well as horticulture, soil science, crop science and agronomy. I hope that it will be a valuable reference for professionals who work with micro irrigation and water management; for professional training institutes, technical agricultural centers, irrigation centers, Agricultural Extension Services, and other agencies that work with micro irrigation programs. I cannot guarantee the information in this book series will be enough for all situations. One must consult an irrigation engineer for an optimum design.

After my first textbook on *Drip/Trickle or Micro Irrigation Management* by Apple Academic Press Inc., and response from international readers, Apple Academic Press Inc. has published for the world community the ten-volume series on *Research Advances in Sustainable Micro Irrigation*, edited by Megh R. Goyal. To get more details on these 10-book volumes, please visit www.appleacademicpress.com.

This book volume is part of the book series *Innovations and Challenges in Micro Irrigation*. This book volume includes reports and studies on technologies to estimate evapotranspiration and to evaluate parameters that are needed in the management of micro irrigation, with worldwide applicability to irrigation management in agriculture. Both book series are musts for those interested in irrigation planning and management, namely, researchers, scientists, educators and students.

The contributions by the cooperating authors to this book volume have been most valuable in the compilation. Their names are mentioned in each chapter and in the list of contributors. This book would not have been written without the valuable cooperation of these investigators; many of them are renowned scientists who have worked in the field of micro irrigation throughout their professional careers.

I will like to thank editorial staff, Sandy Jones Sickels, Vice President, and Ashish Kumar, Publisher and President at Apple Academic Press, Inc., for making every effort to publish the book when the diminishing water resources are a major issue worldwide. Special thanks are also due to the AAP production staff. We request that the reader offer us your constructive suggestions that may help to improve the future editions.

I express my deep admiration to my family for their understanding and collaboration during the preparation of this book, especially my wife Subhadra Devi Goyal. With my whole heart and best affection, I dedicate this book to her, who has taught me patience, perseverance and love for humanity. My salute to her for her social legacy.

As an educator, there is a piece of advice to one and all in the world: *"Permit that our almighty God, our Creator and excellent Teacher, irrigate the life with His Grace of rain trickle by trickle, because our life must continue trickling on..."*

—Megh R. Goyal, PhD, PE, Senior Editor-in-Chief

WARNING/DISCLAIMER

The goal of this compendium, **Performance Evaluation of Micro Irrigation Management**, is to guide the world community on how to manage efficiently for economical crop production. The reader must be aware that dedication, commitment, honesty, and sincerity are the most important factors in a dynamic manner for complete success. This reference is not intended for a one-time reading; we advise you to consult it frequently. To err is human. However, we must do our best. Always, there is a place for learning from new experiences.

The editor, the contributing authors, the publisher, and the printer have made every effort to make this book as complete and as accurate as possible. However, there still may be grammatical errors or mistakes in the content or typography. Therefore, the contents in this book should be considered as a general guide and not a complete solution to address any specific situation in irrigation. For example, one size of irrigation pump does not fit all sizes of agricultural land and will not work for all crops.

The editor, the contributing authors, the publisher, and the printer shall have neither liability nor responsibility to any person, organization, or entity with respect to any loss or damage caused, or alleged to have caused, directly or indirectly, by information or advice contained in this book. Therefore, the purchaser/reader must assume full responsibility for the use of the book or the information therein.

The mention of commercial brands and trade names are only for technical purposes and does not imply endorsement. The editor, contributing authors, educational institutions, and the publisher do not have any preference for a particular product.

All weblinks that are mentioned in this book were active on December 31, 2015. The editors, the contributing authors, the publisher, and the printing company shall have neither liability nor responsibility if any of the weblinks are inactive at the time of reading of this book.

ABOUT SENIOR EDITOR-IN-CHIEF

 Megh R. Goyal, PhD, PE, is a Retired Professor in Agricultural and Biomedical Engineering from the General Engineering Department in the College of Engineering at University of Puerto Rico–Mayaguez Campus; and Senior Acquisitions Editor and Senior Technical Editor-in-Chief in Agriculture and Biomedical Engineering for Apple Academic Press Inc. He received his BSc degree in engineering in 1971 from Punjab Agricultural University, Ludhiana, India; his MSc degree in 1977 and PhD degree in 1979 from the Ohio State University, Columbus; and his Master of Divinity degree in 2001 from Puerto Rico Evangelical Seminary, Hato Rey, Puerto Rico, USA. He spent one-year sabbatical leave in 2002–2003 at the Biomedical Engineering Department at Florida International University in Miami, Florida, USA. Since 1971, he has worked as Soil Conservation Inspector (1971); Research Assistant at Haryana Agricultural University (1972–75) and Ohio State University (1975–79); Research Agricultural Engineer/Professor at the Department of Agricultural Engineering of UPRM (1979–1997); and Professor in Agricultural and Biomedical Engineering in the General Engineering Department of UPRM (1997–2012).

He was first agricultural engineer to receive the professional license in Agricultural Engineering in 1986 from College of Engineers and Surveyors of Puerto Rico. On September 16, 2005, he was proclaimed as "Father of Irrigation Engineering in Puerto Rico for the twentieth century" by the ASABE, Puerto Rico Section, for his pioneer work on micro irrigation, evapotranspiration, agroclimatology, and soil and water engineering. During his professional career of 45 years, he has received awards such as Scientist of the Year, Blue Ribbon Extension Award, Research Paper Award, Nolan Mitchell Young Extension Worker Award, Agricultural Engineer of the Year, Citations by Mayors of Juana Diaz and Ponce,

Membership Grand Prize for ASAE Campaign, Felix Castro Rodriguez Academic Excellence, Rashtrya Ratan Award and Bharat Excellence Award and Gold Medal, Domingo Marrero Navarro Prize, Adopted Son of Moca, Irrigation Protagonist of UPRM, and Man of Drip Irrigation by Mayor of Municipalities of Mayaguez/Caguas/Ponce and Senate/Secretary of Agriculture of ELA, Puerto Rico.

He has authored more than 200 journal articles and textbooks, including *Elements of Agroclimatology* (Spanish) by UNISARC, Colombia, and two *Bibliographies on Drip Irrigation*. Apple Academic Press Inc. (AAP) has published several of his books, including *Management of Drip/Trickle or Micro Irrigation, Evapotranspiration: Principles and Applications for Water Management, Sustainable Micro Irrigation Design Systems for Agricultural Crops: Practices and Theory*, among others. Dr. Goyal was the senior editor-in-chief of the 10-volume book series Research Advances in Sustainable Micro Irrigation with Apple Academic Press and is currently editing two book series, Innovations and Challenges in Micro Irrigation and Innovations in Agricultural & Biological Engineering.

BOOK ENDORSEMENTS

I congratulate the editors on the completion and publication of these book volumes under book series on *Innovations and Challenges in Micro Irrigation*. Water for food production is clearly one of the *Grand Challenges of the 21st Century*. Hopefully this book series will help irrigators and famers around the world to increase the adoption of water savings technology such as micro irrigation. I have known Dr. Goyal since 1982.

—Vincent F. Bralts, PhD, PE
Professor and Ex-Associate Dean
Agricultural and Biological Engineering Department
Purdue University, West Lafayette, Indiana

This textbook is user-friendly and is a must for all irrigation planners to minimize the problem of water scarcity worldwide. *Father of Irrigation Engineering in Puerto Rico of 21st Century and World Pioneer on Micro Irrigation*, Dr. Goyal [my longtime colleague] has done an extraordinary job in the presentation of this book series.

—Miguel A Muñoz, PhD
Ex-President of University of Puerto Rico; and Professor/Soil Scientist

I am moved by recalling my association with Dr. Megh Raj Goyal while at Punjab Agricultural University, India. I congratulate him on his professional contributions and the distinction in irrigation. I believe that this innovative book series on micro irrigation will aid the irrigation fraternity throughout the world.

—A. M. Michael, PhD
Former Professor/Director, Water Technology Centre – IARI
Ex-Vice-Chancellor, Kerala Agricultural University, Trichur, Kerala

In providing these resources in micro irrigation, Megh Raj Goyal, as well as the Apple Academic Press, is rendering an important service to irrigators. Dr. Goyal, *Father of Irrigation Engineering in Puerto Rico*, and his colleagues have done an unselfish job in the presentation of this book volume that is thorough and informative. I have known Megh Raj since 1973 when we were working together at Haryana Agricultural University on an ICAR research project in "Cotton Mechanization in India."

—Gajendra Singh, PhD
Ex-President (2010–12) of ISAE, Former Vice Chancellor,
Doon University, Dehradun, India, Former Deputy Director General
(Engineering) of ICAR, and Former Vice-President/Dean/Professor and
Chairman, Asian Institute of Technology, Thailand

Water is becoming increasingly a scarce resource and limiting agricultural development in many developing and developed economies across the world. Developing infrastructure for the water resources development, conservation and management has been the common policy agenda in many economies. The water use efficiency in the agricultural sector, which still consumes over 80% of water, is only in the range of 30–40% in India, indicating that there is considerable scope for improving the existing water use efficiency. Therefore it is necessary to efficiently utilize water to bring additional areas under irrigation so as to reduce the cost of irrigation and increase the productivity per unit area and unit quantum of water. Micro irrigation, particularly drip and sprinkler irrigation, is followed in many developed countries such as the USA, Austria, Germany, Israel, and Great Britain. It is in this context, the present book series by Dr. Megh R. Goyal serves a critical and timely challenge. I sincerely hope that this book series well read across the globe. This book would be very useful for researchers, scholars, and development personnel, commercial firms dealing with micro irrigation equipments, non-government organizations, and policy makers.

—D. Suresh Kumar, PhD
Professor in Agricultural Economics,
Tamil Nadu Agricultural University

Irrigation has been a vital resource in farming since the evolution of humans. Sustained availability of water cannot be possible in the future, and there are several reports across the globe that severe water scarcity might hamper farm production. Hence, in modern-day farming, the most limiting input being water, much attention is needed for conservation and judicious use of the irrigation water for sustaining the productivity of food and other cash crops. Though the availability of information on micro irrigation is adequate, its application strategies must be expanded for the larger benefit of the water-saving technology by clients. I wish the editors of this book series success in all their endeavors, for helping the users of micro irrigation.

—B. J. Pandian, PhD
Dean and Professor, College of Agricultural Engineering,
Tamil Nadu Agricultural University

OTHER BOOKS ON MICRO IRRIGATION TECHNOLOGY BY APPLE ACADEMIC PRESS, INC.

Management of Drip/Trickle or Micro Irrigation
Megh R. Goyal, PhD, PE, Senior Editor-in-Chief

Evapotranspiration: Principles and Applications for Water Management
Megh R. Goyal, PhD, PE, and Eric W. Harmsen, Editors

Book Series: Research Advances in Sustainable Micro Irrigation
Senior Editor-in-Chief: Megh R. Goyal, PhD, PE
Volume 1: Sustainable Micro Irrigation: Principles and Practices
Volume 2: Sustainable Practices in Surface and Subsurface Micro Irrigation
Volume 3: Sustainable Micro Irrigation Management for Trees and Vines
Volume 4: Management, Performance, and Applications of Micro Irrigation Systems
Volume 5: Applications of Furrow and Micro Irrigation in Arid and Semi-Arid Regions
Volume 6: Best Management Practices for Drip Irrigated Crops
Volume 7: Closed Circuit Micro Irrigation Design: Theory and Applications
Volume 8: Wastewater Management for Irrigation: Principles and Practices
Volume 9: Water and Fertigation Management in Micro Irrigation
Volume 10: Innovation in Micro Irrigation Technology

Book Series: Innovations and Challenges in Micro Irrigation
Senior Editor-in-Chief: Megh R. Goyal, PhD, PE

- Principles and Management of Clogging in Micro Irrigation
- Sustainable Micro Irrigation Design Systems for Agricultural Crops: Methods and Practices
- Performance Evaluation of Micro Irrigation Management: Principles and Practices
- Potential Use of Solar Energy and Emerging Technologies in Micro Irrigation
- Micro Irrigation Management: Technological Advances and Their Applications
- Micro Irrigation Engineering for Horticultural Crops: Policy Options, Scheduling, and Design
- Micro Irrigation Scheduling and Practices

PART I

PRINCIPLES OF MICRO IRRIGATION

CHAPTER 1

RECENT EVAPOTRANSPIRATION RESEARCH IN PUERTO RICO

ERIC W. HARMSEN, VICTOR J. REVENTOS, and
JOHN MECIKALSKI

CONTENTS

1.1   INTRODUCTION

Knowledge of evapotranspiration (ET) is essential for efficient agricultural water management. Efficient use of water for agricultural production has become a moral and ethical issue, considering such factors as increasing global population, dwindling water supplies, widespread degraded water quality, climate variability, and 70% of all water withdrawn is used for agriculture, and most of this is used for irrigation [22, 53]. The purpose of this chapter is to summarize recent ET studies conducted in Puerto Rico. Each of these studies has had at its core, the goal of more accurately estimating crop water requirements and increasing water use efficiency.

Puerto Rico is located in the Greater Antilles between the islands of Hispaniola and the U.S. Virgin Islands and has a land area of approximately 9,100 square kilometers. The climate varies significantly over the island. Rainfall is highly influenced by the Eastern Trades Winds and the orographic effect of the Cordillera Central, a chain of east-west-oriented mountains located along the center of the island. Elevations vary from 0 m mean sea level (msl) along the coasts to approximately 1,300 m msl at Cerro de Punta. Annual rainfall varies from 735 mm at Ponce in Southwest Puerto Rico, to 2,160 mm at Mayaguez in Western Puerto Rico, to 4,370 mm at El Yunque National Forest in Northeast (Pico del Este), Puerto Rico. Puerto Rico has wet and dry seasons. The dry season is from December to April and is caused by a low-level temperature inversion in the easterly trade winds [4]. The wet season is generally from May through October, with some reduction in rainfall during the Caribbean mid-summer drought [5]. Angeles et al. [3] provided insight into the influence of Saharan dust and high level wind shear on the rainfall reduction during mid-summer in Puerto Rico. In much of Puerto Rico, it is difficult to establish a new crop during the dry season without irrigation, while in southern Puerto Rico, irrigation is essentially mandatory for crop production because the annual rainfall is only half of the potential ET.

1.2 EARLY EVAPOTRANSPIRATION STUDIES IN PUERTO RICO

Numerous ET studies have been conducted in Puerto Rico over the years. Harmsen [27] presented a review of ET studies in Puerto Rico prior to 2000, while Goyal and Harmsen [21, 34] reported on several additional studies that have occurred since 2000.

Early efforts to determine crop water requirements in Puerto Rico (1950–1980), by necessity, relied on field measurements based on nonweighing lysimeters or soil water balance methods [10: sugarcane; 51: guinea grass, para grass and guinea grass-kudzu and para grass-kudzu mixtures; 52: sugarcane; 1: plantains; 48: rice]. During the 1980s, meteorological methods were employed to estimate crop water requirements. A number of studies were conducted for various crops using the Blaney Criddle [50] method [13: fifteen different vegetable crops]. See also Goyal and González [15; 14: papaya], Goyal and González-Fuentes [20: sugarcane; 16: sorghum; 18: plantain].

The Hargreaves-Samani [HS, 25] reference ET (ETo) method was employed by González-Fuentes and Goyal [12], in combination with a crop coefficient, to estimate the consumptive use for corn. The HS ETo has been estimated at various locations in Puerto Rico, including: Central Aguirre, Fortuna and Lajas substations [16], Vieques Island [17], and at thirty-four separate locations in Puerto Rico in one study alone [19].

A number of pan evaporation studies were conducted to estimate ETo by Goenaga and his colleagues at USDA-TARS – Mayaguez [1993 – plantains under semiarid conditions, 1994 tanier, 1995 – bananas under semiarid conditions, 1998 – banana under mountain conditions] and Santana Vargas [2000 – watermelon under semiarid conditions].

1.3 RECENT EVAPOTRANSPIRATION STUDIES IN PUERTO RICO

Harmsen and Torres-Justiniano [29] compared estimates from the Penman-Monteith ETo method [PM, 2], based on estimated climate data [30], with estimates of HS ETo for thirty-four locations in Puerto Rico and found reasonably good agreement between the two methods. A user-friendly computer program, available to the public, was developed for the climate parameter estimation procedure called PRET [28].

Harmsen et al. [31] evaluated pan coefficient data for evaporation pans, derived by González and Goyal [11]. The objective of the study was to compare pan coefficients, based on pan evaporation data from 1960–1980 with pan coefficients based on pan evaporation data from 1980–2000. The pan coefficient is derived from the equation: kp = E/ETo, where kp is the pan coefficient, E is pan evaporation. In the González and Goyal [11] kp study, they used the Blaney-Criddle method to estimate ETo. The study of Harmsen et al. [31] concluded that there were significant differences in kp values between the two periods, and presented recommendations for new kp values, based on pan data from the later time period (1980–2000) and use of the Penman-Monteith ETo.

Harmsen et al. [33] developed a field methodology for estimating actual ET. The goal of the project was to develop an instrument that provided accurate ET but at a much lower cost than eddy covariance or weighing lysimeter systems. The "ET station" consisted of a movable temperature and humidity sensor raised and lowered between two vertical positions at two-minute intervals (twelve readings at each position) to obtain the humidity gradient. In the theoretical formulation, a humidity gradient flux equation [38] is equated with the generalized Penman Monteith (GPM) equation (equation 3 in 2) and resolved for the bulk surface resistance (r_s). Once r_s is obtained, all parameters and variables were available for estimating ET using the GPM method. The instrument was compared against an eddy covariance system at the University Florida Agricultural Experiment Station near Gainesville Florida in 2004 and at the University of Puerto Rico Agricultural Experiment Station at Lajas, Puerto Rico in 2005. The ET station compared favorably with the Eddie covariance systems. The advantage of the ET station is that its cost is approximately 1/7 the cost of an Eddie covariance system and about 1/20 of the cost of a weighing lysimeter.

Ramirez-Builes [47] conducted several ET studies for common bean (*phaseolus vulgaris*) in Puerto Rico, with topics including: development of linear models for non-destructive leaflet area estimation; physiological response of different common bean genotypes to drought stress; ET and crop coefficients for two common bean genotypes with and without drought stress; surface resistance estimates from micro-meteorological data; crop measurements under variable leaf area index and soil moisture; crop water stress indices and yield components for common bean genotypes in greenhouse and field environments; and water use efficiency and transpiration efficiency for the two common bean genotypes. See also Ramirez et al. [46], Ramirez et al. [45], Porch et al. [41], and Ramirez et al. [43, 44]

Collaboration was initiated in 2009 between the University Puerto Rico and the University of Alabama in Huntsville, resulting in the availability of a remotely sensed solar radiation product for the northern Caribbean region [34]. Solar insolation estimates are developed from GOES visible data at 1 and 2-km resolution over Puerto Rico and the Caribbean, respectively, and are provided at 30-min time frequency each day (~5 am through 8 pm Local Time). The methods of Diak and Gautie [6], Diak et al. [7] and Paech et al. [40] are utilized, with validation of the solar insolation provided in Otkin et al. [39] and Mecikalski et al. [37]. These GOES solar radiation data are a critical input parameter for ETo equations, for example the GPM, the Hargreaves radiation [24] and the Priestly-Taylor [PT; 42] methods, among others. As noted, the spatial resolution of the GOES product is 2-km, however, there is a sub-set of data available for Puerto Rico and the US Virgin islands, which provide 1-km spatial resolution that is critical for obtaining accurate insolation estimates between cumulus convective clouds [40]. The remotely sensed solar radiation product represents a valuable tool for Puerto Rico, because prior to 2009, there were very few solar radiation sensors distributed across the island.

Rojas González [49] evaluated the performance of the Hargreaves radiation equation [24] and Hargreaves-Samani [25] temperature difference equation for estimating ETo under the humid conditions of western Puerto Rico. The daily temperatures were estimated using a lapse rate approach developed by Goyal et al. [23] for Puerto Rico. Authors concluded that ETo is very sensitive to solar radiation and that its estimation using the square root of the temperature difference times the extraterrestrial radiation did a poor job as compared to the estimation of ETo using measured solar radiation. It was suggested that, in the absence of pyranometer data, remotely sensed solar radiation data should be used.

Harmsen et al. [35] developed a geographic information system for reference ET, based on remotely sensed solar radiation, which included estimates of the Penman-Monteith [2], Hargreaves radiation [24] and Priestly and Taylor [42] methods, all based on remotely sensed solar radiation [34]. Currently the algorithm produces daily reference ET for Puerto Rico, Hispaniola and Jamaica. Figure 1.1 shows an example of daily ETo for the three islands for February 1, 2014 (PM method only). Figure 1.1 illustrates a potential problem related to the remotely sensed solar radiation product, for example, a "banding" error in the ETo that was produced by errors in the satellite-derived insolation data (see northern Puerto Rico in Figure 1.1). From January 1, 2009 through the present, the daily ETo images are available at a public website: http://pragwater.com/daily-reference-evapotranspiration-eto-for-puerto-rico-hispaniola-and-jamaica/.

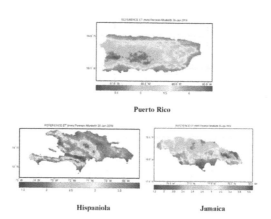

Puerto Rico

Hispaniola   **Jamaica**

FIGURE 1.1 Estimated daily reference evapotranspiration for January 20, 2014 for Puerto Rico, Hispaniola and Jamaica.

Harmsen [26] developed a web-based methodology for irrigators to schedule their irrigation based on the daily ETo estimates for Puerto Rico, Hispaniola and Jamaica. The goal for the methodology is to assist irrigators to replace crop ET (i.e., Kc ETo, where Kc is the crop coefficient) with irrigation (or rainfall) throughout the crop season. In Puerto Rico, the user can access NEXRAD radar rainfall data if they do not have a rain gauge on their farm.

Furthermore, Harmsen et al. [32] modified the algorithm described in the previous paragraph to include the water and energy balance components (the algorithm is called GOES-PRWEB). The water balance is based on the actual ET, which is estimated at 1-km spatial resolution. Actual ET is derived from an energy balance approach similar to the methodology described by Yunhao et al. [54]. Surface runoff is estimated using the Natural Resource Conservation Service Curve Number (CN) method [8]. Rainfall is obtained from the National Oceanic and Atmospheric Administration's (NOAA's) Advanced Hydrologic Prediction Service (AHPS) website. Soil moisture is estimated using a simple soil reservoir concept [34] in which infiltrated water in excess of the field capacity becomes aquifer recharge. Figures 1.2 and 1.3 show examples of the energy and water balance components, respectively, for February 1, 2014. Image data for twenty-five hydro-climate variables are available on a public website: http://pragwater.com/goes-puerto-rico-water-and-energy-balance-goes-web-algorithm/. Archived images are available from January 2009 through the present.

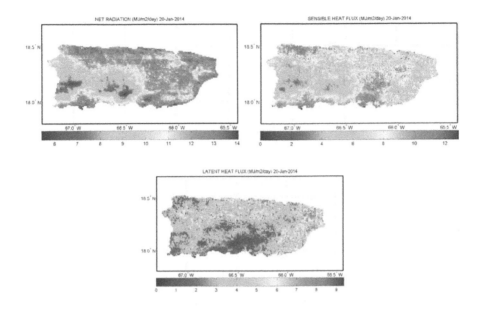

FIGURE 1.2 Daily energy balance components produced by GOES-PRWEB for January 20, 2014.

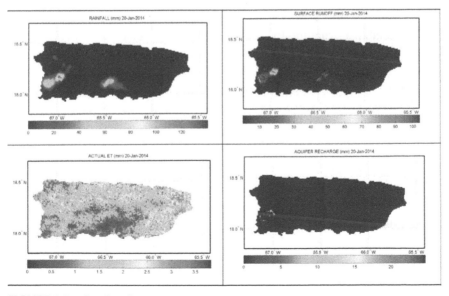

FIGURE 1.3   Continued

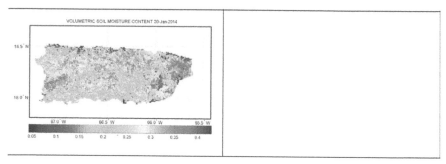

FIGURE 1.3 Daily water balance components produced by GOES-PRWEB for January 20, 2014.

1.4 FUTURE WORK

Current and future efforts on several of the projects are listed below.

a. A mobile/desktop application is being developed for the web-based irrigation scheduling procedure [27]. The application will allow the user to create an account, which will store their irrigation history in a database.

b. The 25 hydro-climate variables are currently only available as images (jpg format). With an improved user interface and web database structure, the user will be able to select their location on a map and obtain time series data for any variable for any date range. Alternatively, the user will be able to download map data for any date range. This latter option will be useful for modelers who may want to use map data (e.g., soil moisture) for initializing model simulations.

c. Currently the surface runoff, aquifer recharge and total stream flow components of GOES-PRWEB are being calibrated using US Geological Survey (USGS) stream gauge data.

d. Rojas (2012) showed that the bias-corrected AHPS rainfall data have a low accuracy in western Puerto Rico. We plan to develop a real-time (daily) bias correction factor for each of the USGS gauged watersheds in Puerto Rico. The correction fac-tor will be derived using a methodology similar to Fekete and Vorosmarty [9].

1.5 CONCLUSIONS

This chapter summarizes recent ET studies in Puerto Rico. Studies included the development of procedures for estimating climate parameters for use in the PM ETo equation and a computer program for implementing the method; recalculation of evaporation pan coefficients for estimating ETo, development of the field instrument for measuring actual ET in the field, and algorithm for estimating reference ET in Puerto Rico, Hispaniola and Jamaica, a irrigation scheduling procedure which utilized the ETo map data for the three islands, and an algorithm that produces a daily water and energy balance at 1-km spatial resolution in Puerto Rico.

1.6 SUMMARY

Since 2000, several evapotranspiration (ET) research projects have been conducted in Puerto Rico. These projects involved the development of procedures for estimating weather variables for use in the Penman-Monteith ET equation, design and testing of field equipment for measuring actual ET, derivation of crop coefficients and the application of satellite remote sensing techniques for estimating reference and actual ET. The climate parameter estimation project resulted in a software product called PR-ET, which allows the user to estimate the reference and crop evapotranspiration for any location in Puerto Rico with only day-of-year, site latitude and elevation. An "ET-Station" was designed and compared to eddy-covariance systems in Florida and Puerto Rico. The new equipment performed well and could be produced for a fraction of the cost of traditional field ET measurement equipment. A three-year study employed the ET-Station in southern Puerto Rico where crop coefficients were derived for three varieties of common bean. Recently, a research project was conducted in which an operational water and energy balance algorithm called GOES-PRWEB was developed based on 1-km resolution satellite-derived solar radiation. GOES-PRWEB data are published daily for Puerto Rico, and can be used by the scientific community, private companies and citizens, and the government. Reference ET and the related input variables are also published each day for the islands of Hispaniola and Jamaica.

Recent outreach efforts have resulted in an online procedure for scheduling irrigation at any location throughout the three islands.

ACKNOWLEDGEMENTS

This research received partial funding from the following sources: NOAA-CREST (Grant NA06OAR4810162) and USDA Hatch Project (Hatch-402).

KEYWORDS

- ASAE
- Caribbean
- Dominican Republic
- Evapotranspiration
- GOES-PRWEB
- Haiti
- Hargreaves-Samani
- Hispaniola
- Jamaica
- latent heat
- Penman-Monteith
- Priestley-Taylor
- Puerto Rico
- reference evapotranspiration
- satellite remote sensing
- solar radiation

REFERENCES

1. Abruña, F., Vicente-Chandler, J., Irizarry, H., & Silva, S. (1979). Evapotranspiration with plantains and the effect of frequency of irrigation on yields. *J. Agric. U.P.R.*, 64(2), 204–210.

2. Allen, R. G., Pereira, L. S., & Dirk Raes, Smith, M. (1998). *Crop Evapotranspiration Guidelines for Computing Crop Water Requirements.* FAO Irrigation and Drainage Paper 56, Food and Agriculture Organization of the United Nations, Rome. 300 pp.
3. Angeles, M. E., González, J. E., Ramírez–Beltrán, N. D., Tepley, C. A., & Comarazamy, D. E. (2010). Origins of the Caribbean rainfall bimodal behavior. *J. Geophys. Res.,* 115(D111060), 1–17.
4. Capiel, M., & Calvesbert, R. J. (1976). On the climate of Puerto Rico and its agricultural water balance. *J. Agric. Univ. P. R.,* 60(2), 139–153.
5. Curtis, S., & Gamble, D. W. (2007). Regional variations of the Caribbean midsummer drought. *Theor. Appl. Climatol.,* doi: 10.1007/s00704-007-0342-0.
6. Diak, G. R., & Gautier, C. (1983). Improvements to a simple physical model to estimate insolation from GOES data. *J. Clim. Appl. Meteorol.,* 22, 505–508.
7. Diak, G. R., Bland, W. L., & Mecikalski, J. R. (1996). A note on first estimates of surface insolation from GOES-8 visible satellite data. *Ag. For. Meteor.,* 82, 219–226.
8. Fangmeier, D. D., Elliot, W. J., Workman, S. R., Huffman, R. L., & Schwab, G. O. (2005). *Soil and Water Conservation Engineering.* Fifth Edition. 502 pp.
9. Fekete, B., & Charles, Vorosmarty, J. (2002). High-resolution fields of global runoff combining observed river discharge and simulated water balances. *Global Biochemical Cycles,* 16(3), 15–16.
10. Fuhriman, D. K., & Smith, M. (1951). Conservation and consumptive use of water with sugarcane under irrigation in the south coastal area of Puerto Rico. *J. of Agric. University of Puerto Rico,* 35(1), 1–45.
11. González, E., & Goyal, M. R. (1989). Coeficiente de Evaporímetro para Puerto Rico. *J. Agric. Univ. P. R.,* 73(1).
12. Gonzalez-Fuentes, E. A., & Goyal, M. R. (1988). Estimación de los requisitos de riego para maíz dulce (*Zea Mays cv.* Suresweet) en la costa sur de Puerto Rico. *Journal of Agriculture of the University of Puerto Rico,* 72(2), 277–284.
13. Goyal, M. R. (1989a). *Estimation of Monthly Water Consumption by Selected Vegetable Crops in the Semiarid and Humid Regions of Puerto Rico.* AES Monograph 99–90, June, Agricultural Experiment Station, University of Puerto Rico Rio Piedras, PR.
14. Goyal, M. R. (1989b). Research Note: Monthly consumptive use of papaya at seven regional sites in Puerto Rico. *J. Agric. U.P.R.,* 73(2).
15. Goyal, M. R., & González, E. A. (1988). Requisitos de riego para sorgo en las costas Sur y Norte de Puerto Rico. *J. Agric. University of Puerto Rico,* 72(4), 585–598.
16. Goyal, M. R. (1988). Research Note. Potential evapotranspiration for the south coast of Puerto Rico with the Hargreaves-Samani technique. *J. Agric. U.P.R.,* 72(1), 57–63.
17. Goyal, M. R. (1988). Potential Evapotranspiration for Vieques Island, Puerto Rico, with the Hargreaves and Samani Model, Research Note. *Journal of Agriculture, U.P.R.,* 72(1), 177–178.
18. Goyal, M. R., & González, E. A. (1988). Requisitos de riego para plátano en siete regions ecológicas de Puerto Rico. *J. Agric. University of Puerto Rico,* 72(4), 599–608.
19. Goyal, M. R., & González, E. A. (1988). Water Requirements for Vegetable Production in Puerto Rico. *ASCE Symposium on Irrigation and Drainage,* July.
20. Goyal, M. R., & González-Fuentes, E. A. (1989). Estimating water consumptive use by sugarcane at four regional sites in Puerto Rico. *J. Agric. U.P.R.,* 73(1).

21. Goyal, M. R., & Harmsen, E. W. (eds.), (2014). *Evapotranspiration Principles and Applications in Water Management*. Oakville, ON, Canada: Apple Academic Press Inc.
22. Goyal, M. R. (ed.), (2015). *Research Advances in Sustainable Micro Irrigation, volumes 1 to 10*. Oakville, ON, Canada: Apple Academic Press Inc.,
23. Goyal, M. R., Gonzalez, E. A., & Chao de Baez, C. (1988). Temperature versus elevation relationships for Puerto Rico. *Journal of Agriculture, U.P.R.*, 1988–72(3), 440–467.
24. Hargreaves, G. H. (1975). Moisture availability and crop production, *Transactions of the ASAE*, 18(5), 980–984.
25. Hargreaves, G. H., & Samani, Z. A. (1985). Crop evapotranspiration from temperature. *Appl. Eng. Agric. of ASAE*, 1(2), 96–99.
26. Harmsen, E. W. (2012). Technical Note: A Simple Web-Based Method for Scheduling Irrigation in Puerto Rico. *J. Agric. Univ. P. R.* 96, 3–4.
27. Harmsen, E. W. (2003). Fifty years of crop evapotranspiration studies in Puerto Rico. *Journal of Soil and Water Conservation*. July/August, 58(4), 214–223.
28. Harmsen, E. W., & Gonzaléz, A. (2002). *Puerto Rico Evapotranspiration Estimation Computer Program PR-ET Version 1.0 USER'S MANUAL*. Prepared for the University of Puerto Rico Experiment Station-Rio Piedras. Grant SP-347. August (2002).
29. Harmsen, E. W., & Torres-Justiniano, S. (2001). Estimating island-wide reference evapotranspiration's for Puerto Rico using the Penman-Monteith method. ASAE Paper 01–2174. 2001 American Society of Agricultural Engineering Annual International Meeting. Sacramento Convention Center, Sacramento, CA, USA. July 30-August 1.
30. Harmsen, E. W., & Torres-Justiniano, S. (2001). Evaluation of prediction methods for estimating climate data to be used with the Penman-Monteith equation in Puerto Rico. ASAE Paper 01–2048. American Society of Agricultural Engineers Annual International Meeting.
31. Harmsen, E. W., Gonzaléz, A., & Winter, A. (2004). Re-evaluation of pan evaporation coefficients at seven locations in Puerto Rico. *J. Agric. Univ. P. R.*, 88(3–4), 109–122.
32. Harmsen, E. W., Mecikalski, J., Mercado, A., & Tosado Cruz, P. (2010). Estimating evapotranspiration in the Caribbean Region using satellite remote sensing. Proceedings of the AWRA Summer Specialty Conference, Tropical Hydrology and Sustainable Water Resources in a Changing Climate. San Juan, Puerto Rico. August 30–September 1.
33. Harmsen, E. W., Mecikalski, J., Cardona-Soto, M. J., Rojas Gonzalez, A., & Vasquez, R. (2009). Estimating daily evapotranspiration in Puerto Rico using satellite remote sensing. *WSEAS Transactions on Environment and Development*, 6(5), 456–465.
34. Harmsen, E. W., Trinidad-Colon, J., Arcelay, C. L., & Rodriguez, D. C. (2014). Irrigation scheduling for sweet pepper. Chapter 29, In: *Evapotranspiration Principles and Applications in Water Management* by, M. Goyal, and E. W. Harmsen (eds.), Apple Academic Press Inc., 588 pp.
35. Harmsen, E. W., Ramirez-Builes, V. H., Dukes, M. D., Jia, X., Gonzalez, J. E., & Pérez Alegría, L. R. (2009). A ground-based method for calibrating remotely sensed surface temperature for use in estimating evapotranspiration. *WSEAS Transactions On Environment and Development*, 1(5), 13–23.

36. Harmsen, E. W., Tosado, P., & Mecikalski, J. (2014). Calibration of selected pyranometers and satellite derived solar radiation in Puerto Rico. *International Journal of Renewable Energy and Technology*, 5(1), 43–54.

37. Mecikalski, J. R., Sumner, D. M., Jacobs, J. M., Pathak, C. S., Paech, S. J., & Douglas, E. M. (2011). Use of visible geostationary operational meteorological satellite imagery in mapping reference and potential evapotranspiration over Florida. In: *Evapotranspiration.* ISBN 978-953-307-251-7, Editor Leszek Labedzki, Chapter 10, pgs. 229–254.

38. Monteith, J. L., & Unsworth, M. H. (2008). *Principles of Environmental Physics.* 3rd ed., Academic Press. 418 pgs.

39. Otkin, J. A., Anderson, M. C., Diak, G. R., & Mecikalski, J. R. (2005). Validation of GOES-based insolation estimates using data from the United States climate reference network. *Hydrometeor, J.,* 6, 460–475.

40. Paech, S. J., Mecikalski, J. R., Sumner, D. M., Pathak, C. S., Wu, Q., Islam, S., & Sangoyomi, T. (2009). Satellite-based solar radiation in support of potential and reference evapotranspiration estimates over Florida: A 10-year climatology. *J. Amer. Water Res. Assoc.*, 45(6), 1328–1342.

41. Porch, G, T., Ramirez, V. H., Santana, D., & Harmsen, E. W. (2009). Evaluation of drought tolerance in common bean germplasm in Juana Díaz, Puerto Rico. *Journal of Agronomy and Crop Science,* 195, 1–7.

42. Priestly, C. H. B., & Taylor, R. J. (1972). On the assessment of surface heat flux and evaporation using large scale parameters. *Monthly Weather Review*, 100, 81–92.

43. Ramirez, B., V. H., Porch, T. G., & Harmsen, E. W. (2008). Evaporation and water use efficiency for common bean genotypes under non-stress and drought stress conditions. *Annual Report of the Bean Improvement Cooperative*, 51, 82–83.

44. Ramirez-Builes, V. H., Harmsen, E. W., Porch, T. G. (2008). Estimation of actual evapotranspiration using measured and calculated values of bulk surface resistance. *Proceedings of the ASCE World Environmental and Water Resources Congress.* May 13–16, Honolulu, Hawaii.

45. Ramirez, V. H., Porch, T. G., & Harmsen, E. W. (2006). Effects of drought on stomatal resistance, surface resistance and leaf temperature in four common bean genotypes (Phaseolus vulgaris, L.). *Proceedings of the 42nd Annual Meeting of the Caribbean Food Crops Society*, Vol. XLII-Number 2. July 9–14, San Juan, Puerto Rico.

46. Ramirez, V. H., Porch, T. G., & Harmsen, E. W. (2011). Genotypic differences in water use efficiency of common bean under drought stress. *Agronomy Journal*, 103(4), 1206–1215.

47. Ramírez-Builes, Víctor, H. (2007). *Plant- Water Relationships for Several Common Bean Genotypes (Phaseolus vulgaris, L.) with and Without Drought Stress Conditions.* MS Thesis. Department of Agronomy and Soils, University of Puerto Rico – Mayaguez, Puerto Rico, December.

48. Ravalo, E. J., & Goyal, M. R. (1988). Water requirement of rice in Lajas Valley, Puerto Rico. Dimension, Año 2, Vol. 8, Enero-Febrero-Marzo.

49. Rojas González, A. M. (2012). Flood Prediction limitations in small watersheds with mountainous terrain and high rainfall variability. PhD Dissertation, Department of Civil Engineering and Surveying, University of Puerto Rico – Mayaguez Campus, pp. 200.

50. SCS, (1970). *Irrigation Water Requirements*, Technical Release No. 21. USDA Soil Conservation Service, Engineering Division.
51. Vázquez, R. (1965). Effects of Irrigation and nitrogen levels on the yields of Guinea Grass, Para Grass, and Guinea grass-Kudzu and Para grass-Kudzu in Lajas Valley. *Journal of Agriculture of the University of Puerto Rico*, XLLIX(4), 389–412.
52. Vázquez, R. (1970). *Water Requirements of Sugarcane Under Irrigation in Lajas Valley, Puerto Rico*. Bulletin 224, University of Puerto Rico Agricultural Experiment, Rio Piedras, PR.
53. WBCSD, (2008). *Agriculture and Ecosystems Facts and Trends*. World Business Council for Sustainable Development. July.
54. Yunhao, C., Xiaobing, L., & Peijun, S. (2001). Estimation of regional evapotranspiration over Northwest China by using remotely sensed data. *Journal of Geophysical Sciences*, 11(2), 140–148.

CHAPTER 2

MODELING OF EVAPORATION FROM BARE SOIL

INGLE VISHAL KESHAORAO and A. K. MISHRA

CONTENTS

[a]Edited version of: *Keshaorao, I. V., 2009. Modeling of Evaporation from Bare Soil. M.Tech Thesis (Unpublished). Division of Agricultural Engineering, Indian Agricultural Research Institute, New Delhi – 110012, India.*

[b]Authors acknowledge the financial assistance and fellowship by the Indian Council of Agricultural Research (ICAR) during the research study.

2.1   INTRODUCTION

Evapotranspiration from earth's surface is one of the most important components of the global water cycle. There is a lack of knowledge in quantification of evapotranspiration and its partitioning into components (bare soil evaporation and plant transpiration). Bare soil evaporation is an important water balance component during early growth stages of irrigated field crops, row crops with incomplete cover and in soils with high water table. Quantification of bare soil evaporation can help in irrigation and environmental management. Evaporation is known to be influencing many other hydrologic and ecological processes too.

Evaporation from bare soil surface is the loss of water surrounding the soil particles as thin films (hygroscopic water) and the water filling the pore spaces (soil water) into the atmosphere. Knowledge of evaporation is important in hydrologic water balances, irrigation scheduling, crop yield forecasting, water resource management, irrigation system design and climate change forecasting, etc. The prediction of evaporation and transpiration together is referred to as evapotranspiration, and is required for a reliable project planning, design and operating of an irrigation system in an irrigation command, water resources monitoring, water harvesting and storage of rainwater, and management of catchments for efficient utilization [45].

In arid ecosystems, where vegetation and transpiration is minimized and drainage is low, evaporation can be dominant sink of soil moisture and can influence soil water redistribution. The magnitude of loss of water by evaporation can be of great importance in the fields of land management, liquid waste disposal and ecosystem management [53]. The effect of soil texture and climatic water demand on evaporation dynamics is important for developing a good understanding of ecosystems where evaporation plays an important role in the water balance [15, 27].

The magnitude and rate of removal of water as evaporation is governed by atmospheric evaporativity as well as soil transmission properties. This process results into considerable losses of water in both irrigated and unirrigated agriculture if not checked properly. Evaporation depletes large amounts of soil moisture from soil surface during tillage operations, planting, germination and early seedling growth stages when soil largely remains bare, thus hampering the growth of young plants during their most vulnerable stage. In early vegetative phases of plant growth, when the plants are young and a high leaf area index has not been achieved, evaporation from bare soil result into considerable moisture depletion from the soil [2, 33, 48].

Evaporation from the soil surface especially in semi-arid regions constitutes a large fraction of the total water loss not only from bare soils, but also from cropped fields [34, 74]. In annual field crops, the soil surface remains bare for many weeks resulting into substantial evaporation. It is more prominent during periods of seed germination and seedling establishment as well as during the subsequent growth of the young crop. Evaporation from bare soils results in a considerable loss of moisture and has a direct impact on crop yield in rainfed agriculture of arid and semi-arid regions. In orchards too, the soil surface between the trees is kept bare by frequent tillage and is continuously subjected to evaporation [48].

Transpiration through crops is regarded as a beneficial depletion, but the evaporation from the bare soil in irrigated fields, with a partial canopy cover or from weeds, can be considered as a non-beneficial depletion [19]. The transpiration from the canopy layers and evaporation from the soil can be separately calculated using equations of the Penman–Monteith type [14]. Acs [1] analyzed the bare soil evaporation using different methods and compared the performance of each. In most biophysical models evapotranspiration is split into transpiration and bare soil evaporation [22]. Transpiration is biologically regulated through plant stomata, while soil evaporation occurs through the soil pores.

Under natural field conditions, evaporation from bare soil surfaces is extremely variable in time and space. It strongly depends on both surface soil moisture and meteorological conditions. However, most experimental studies of soil evaporation have been conducted under controlled conditions mainly in the laboratory. Only a few measurements of soil evaporation have been made under natural field conditions over a relatively long period, therefore, the actual state of soil evaporation under such conditions has not yet been fully investigated.

Evaporation from bare soil (Es) is often characterized as occurring in two distinct stages [25, 38, 62–64]. The first stage is termed the "energy limited" stage. During this stage, moisture is available at or is transported to near the soil surface at a rate sufficient to supply the potential rate of evaporation, defined as the rate of evaporation constrained by energy availability at the soil surface. The second stage is termed as the "falling rate stage" or "soil limited stage," where hydraulic transport of subsurface moisture to near the soil surface is unable to supply water at the potential evaporation rate. During falling rate stage, the soil surface appears dry and a portion of the evaporation occurs below the soil surface. The subsurface evaporation is caused by transport of heat from the soil surface into the soil profile [3].

Several simplified models have been advocated for the relation between soil evaporation and water potential. Ehlers et al. [20] and Aydin [5] estimated evaporation from bare soils as a function of soil water potential (matric potential) at the surface layer, neglecting the influence of the hydraulic gradient. Similarly, Beese et al. [8] quantified the relationship between soil evaporation and water potential at the soil surface.

The global average surface air temperature is projected to increase by 1.4–5.8°C over the period of 1990–2100 [29]. Precipitation is not likely to increase in semi-arid regions, where the effects of climate change on soil water balance are of major concern as the increased temperature stimulates the evaporative demand of the atmosphere. In such regions, where summer fallow is practiced, the bare soil evaporation may result into the loss of the most of the incoming precipitation. The soil surface in such areas remains bare for sufficiently long times ranging from days to weeks. In absence of good or sufficient rainfall, soils of these regions experience sever soil moisture deficit. If attempts are made to grow field crops in these areas soon after the onset of first rain, the seed germination, seedling establishment, and/or subsequent growth of the young crop may result into a failure.

In the era of global climate change and consequently the reducing moisture availability, the knowledge of bare soil evaporation is of paramount importance for planning and executing agricultural operations for growing crops. In India too, this phenomenon has got special significance as during the summer season large barren tracts experience intense heat and wind, resulting into a completely dry soil profile. The climatic water demands remains quite high even with the onset of monsoon for few days so much so that almost whole amount of first rain gets soaked and redistributed into the soil profile. Subsequently, this moisture gets evaporated and lost without any beneficial use. A quantitative assessment of bare soil evaporation from the agricultural lands will help the planners to design appropriate strategies for land preparation, sowing and growing crops with less or limited water supply situations.

Hence, the present study was planned with the following specific objectives: (i) To estimate the bare soil evaporation from different soil types. (ii) To develop bare soil evaporation prediction equations. (iii) To validate prediction equation from observed data.

2.2 LITERATURE REVIEW

Soil is the basis of human's living. Soil moisture plays a significant role in studying the matter and energy exchanges in global hydrologic sphere. The evaporation of soil moisture has an influence on the water vapor cycle. Soil moisture is one of the primary measurable parameters in crop yield estimation and water resources management [32]. Evaporation or the net rate of vapor transport to the atmosphere has great importance in many disciplines, including irrigation system design, irrigation scheduling, hydrologic and drainage studies etc.

Evaporation from the bare soil surface is essentially the evaporation of water surrounding the soil particles as thin films (hygroscopic water) and filling the pore spaces between them (soil water). Therefore, the atmospheric conditions that govern the evaporation from the free water surface will also govern the rate of evaporation from the bare soil surface. However, in case of evaporation from the free water surface, the supply of water is not a limiting factor while the evaporation from the bare soil is affected by the insufficient supply of water. Furthermore, the water molecules escaping from the soil will have to overcome greater resistance due to the attraction of soil particles (adhesive forces) than while escaping from a free water surface (cohesive forces). When the water content of the surface soil reduces below a threshold value, evaporation practically ceases to exist.

The above discussion highlights that evaporation of water from the bare soil demands three basic physical requirements to be fulfilled:

(i) A continuous supply of heat to change the state of water from liquid to vapor should be available;
(ii) a vapor pressure gradient between the soil surface and the surrounding atmosphere should be maintained; and
(iii) a continuous supply of water from or through the soil profile at the surface.

The first two conditions, namely, supply of energy and removal of vapor are external to the evaporating body and are influenced by the meteorological factors such as maximum/minimum ambient air temperature, relative humidity, wind velocity and solar radiation, which together determine the evaporative demand of the atmosphere or evaporation potential. The evaporation potential is defined as the maximum evaporation rate from the surface of free water in bulk under the given meteorological conditions. The third condition determines the maximum rate at which soil can transmit water to the plane of evaporation. Thus, the evaporation rate is determined either by the evaporation potential or by the soil's own ability to transmit water to the plane of evaporation under the given condition, whichever is lesser [23].

2.2.1   SOIL EVAPORATION STAGES

Evaporation occurs when water is converted in water vapor. The rate is controlled by availability of energy at the evaporating surface, and the ease with which water vapor can be diffused into the atmosphere. Different physical processes are responsible for the diffusion, but the physics of water vapor loss from open water surface and from soil and crops is essentially identical. The evaporation is defined as the rate of liquid water transformation to vapor from open water, bare soil, or vegetation with soil beneath. Unless otherwise stated, this rate is in millimeter of evaporated water per day. In the case of vegetation growing in soil, transpiration is defined as that part of the total evaporation, which enters the atmosphere from the soil through the plants. The soil evaporation (E_s) is a key component in the water balance especially under dryland farming system where the soil is exposed to the atmosphere for a long time from the beginning of the growing season to the canopy development and maturity in some sparsely growing crops.

The rate of evaporation has traditionally been estimated using metrological data from climate stations located at particular points within region. It has been assumed that the evaporation area is sufficiently small and that the evaporation has no effect on regional climate or air movement. In reality, this simplified approach approximates a more complex situation in which local evaporation is a function of both local climate and regional air movement.

Ritchie [62] described a model for estimating Es and also evaporation from the soil under crop. The model is based on a two-stage evaporation theory in which the "First Stage (E_{S1w})" comes into effect soon after recharge of the soil profile and depends on the "Potential Evaporation Rate (E_{sp})" until reaching a theoretical threshold. The "Second Stage (E_{S2w})" comes into effect afterwards that depends on the moisture content and the hydraulic conductivity of the soil. Therefore, evaporation from bare soil is:

$$\sum E_s = \sum E_{S1w} + \sum E_{S2w} \tag{1}$$

where:

$$\sum E_{S,1} = \sum E_0 \text{ when } \sum E_s \leq U \tag{2}$$

$$\sum E_{S,2} = C\, t^{\frac{1}{2}} \text{ when } \sum E_s > U \tag{3}$$

where, E_0 is the potential evaporation determined with the equation of Priestley and Taylor [57]. The amount of water that can be evaporated from a given soil type during $E_{S,1}$ is denoted by U. Time is denoted by t and C represents a soil parameter which can be expressed using Eq. (4) given by Black et al. [9].

$$C = 2(\theta_i - \theta_0)(D/\pi)^{1/2} \tag{4}$$

where, θ_0 is the moisture content at the soil surface at a particular time during the drying period, θ_i is the initial moisture content at any given depth, and D is the soil water diffusivity.

Black et al. [9] also showed that C can be taken as the slope of the regression of $\sum E_{S,2}$ on the square root of time (t), after the onset of $E_{S,2}$. Studies, however, showed that U and C are sensitive to E_0 [24, 31, 35, 37]. Johns [37] found that U is inversely proportional to E_0 while

Jackson et al. [31] reported an almost linear increase in C with seasonal temperature. Thus, estimates of $\sum E_S$ by the Ritchie's model [62] should depend on the conditions under which these parameters were determined.

Boesten et al. [11] and Stroosnijder [71] proposed an alternative model in which both U and C were replaced by a single parameter β that depended on E_0. The parameter β is similar to the parameter C, but is obtained by regressing $\sum E_{S,2}$ on the square root of $\sum E_0$. According to this model, $\sum E_s$ is estimated as:

$$\sum E_{S,1} = \sum E_{0,} \text{ when } \sum E_0 \leq \beta^2 \qquad (5)$$

$$\sum E_{S,2} = \beta(\sum E_0)^{1/2} \text{ when } \sum E_0 \geq \beta^2 \qquad (6)$$

where, E can be determined from the Penman-Monteith equation [50].

Jalota and Prihar [36] evaluated the effects of atmospheric evaporativity, soil type and redistribution time on evaporation from the bare soil. Experiments were conducted with silt loam, sandy loam and loamy sand soils with 0 and 2 days redistribution time before commencement of evaporation under high (15.1 ± 0.50 mm day^{-1}), medium (10.1 ± 0.50 mm day^{-1}) and low (6.3 ± 0.52 mm day^{-1}) E, to ascertain if cumulative evaporation (CE) was always greater under higher E, irrespective of experimental conditions, and if the evaporation rates during falling rate stage were insensitive to changes in E, in all soils. They concluded that where evaporation commenced immediately after wetting, CE up till 30 days was always higher under higher than lower E, values in the silt loam and sandy loam soils. In the loamy sand, however, CE under medium E, conditions exceeded that under high E, conditions after 2 days. When evaporation commenced after 2 days of redistribution CE under medium and low E, values exceeded that under high E, values after 8 and 12 days in the sandy loam, and 2 and 6 days in loamy sand, respectively. Where evaporation was commenced after the 2-day redistribution, CE at 30 days in the silt loam and sandy loam was 12% less than where evaporation commenced immediately after infiltration under all the E, values. In loamy sand these differences were 17% and 40% under low and high E, values, respectively. CE as well as evaporation rate (ER) were sensitive to E, in the initial period of falling rate stage in the silt loam and the sandy loam but not in the loamy sand.

2.2.2 ESTIMATION OF EVAPORATION

Many approaches are available for measurement of evaporation. The following methods or approaches embody the estimation of evaporation from bare soil surface:

- the aerodynamic method;
- the energy balance method;
- the combination method; and
- the soil water evaporation estimation from the solution of water flow from equations using water transport properties of the soil.

Since the present study confines to the field experiments and soil column studies in the laboratory simulation by following the approach of evaporation measurement using water balance approach, a comprehensive review pertinent to this subject has been presented in the following subsections.

2.2.2.1 Soil Moisture Depletion Methods

Evaporation under field conditions can be determined by measuring the change of soil water content over a period of time. The soil water content can be measured by gravimetric sampling, tensiometers, gypsum blocks, neutron moisture meter, gamma attenuation technique, pressure transducers, time domain reflectometry (TDR), and frequency domain reflectometry (FDR) or by using infrared thermometers etc. Other methods used for measuring soil water content include tensiometer and gypsum blocks. These techniques require calibration to determine the amount of water that must be applied to refill the profile.

Deki et al. [16] studied the sensitivity of bare soil evaporation schemes to soil surface wetness, using the coupled soil moisture and surface temperature prediction model. The performance of evaporation schemes with α and β approach and their combination within resistance representation of evaporation from bare soil surface was discussed. For this purpose nine schemes, based on different functions of α or β on the ratio of the volumetric soil moisture content and its saturated value were used. A sensitivity analysis was made using two sets of data derived from the volumetric soil moisture content of the topsoil layer. One with values below the wilting point (0.17 m^3m^{-3}) and the second with values above 0.20 m^3m^{-3}.

Ventura et al. [73] used continuous soil moisture measurements and hourly reference evapotranspiration data to estimate a soil hydraulic factor (β) for modeling soil evaporation. A two stage soil evaporation model was proposed using the β factor by assessing the end of the energy limited soil evaporation phase (Stage 1) and the evaporation rate during the soil hydraulic limited phase (Stage 2) which uses a hydro-probe soil moisture measuring device to estimate the continuous soil evaporation. The estimation of evaporation with soil moisture sensors was simpler and less expensive as compared to the energy balance technique. When daily soil evaporation from soil moisture measurements was compared with soil evaporation estimated from energy balance measurements, the root-mean-square error was 1.3 mm day^{-1}. Direct soil monitoring method had bigger error, but the method is less costly.

2.2.2.2   Lysimeter Studies

Singh, et al. [66] has discussed the design requirements for installation and proper use of lysimeters. A lysimeter is a container located in hydrologically isolated conditions from its surroundings, which contains a volume of soil; with or without crop. For accurate and reliable measurement of evapotranspiration the lysimeter should be constructed [77], installed and operated properly. Weighing type lysimeters are preferred over the nonweighing type ones due to the accuracy of measurements.

Boast and Robertson [10] used a "micro-lysimeter" for estimating evaporation from soil. It consisted of a thin-walled cylinder of 76 mm in diameter. The unit was pushed into the field soil. The soil-filled cylinder was removed from the field and made watertight by closing at the bottom. Thus, by determining the mass of the micro-lysimeter, replacing it in the field with its top surface even with the surrounding soil, leaving it exposed to environmental conditions for a period of time (typically 1 day), and re-determining its mass. Evaporation loss from the micro-lysimeter is the difference between the two masses. The deviations are quantified by comparing short micro-lysimeters with effectively "infinitely long" ones. The deviation is influenced by the time that the micro-lysimeter is exposed to environmental conditions and by the length of the soil sample.

For a silty clay loam soil under evaporativity conditions ranging from 2 to 9 mm/d, the measurement error for micro-lysimeter of 70 mm in length was found to be <0.5 mm for 1 or 2 days, depending on whether the initial soil condition was "wet" or "dry," respectively (0.26 or 0.13 g of H_2O per gram of dry soil in the top 20 mm). Hence, for many applications the method is valid for 1 or 2 days. Correction equations for deviations of up to 0.5 mm were also given. This method required little instrumentation and made it possible to measure the evaporation under field conditions (e.g., at large numbers of sites or in areas of partial crop shading) where micrometeorological, water balance, and traditional lysimeter methods were unpractical or impossible [77].

Jackson et al. [30] studied evaporation from soil (E_s) in an agroforestry system using soil micro-lysimeter, where the trees and crops were grown, in monoculture, two components grown separately, and in bare soil at an equatorial field site in Kenya. E_s varied according to the different shade regimes and as a function of proximity to trees and/or crops. Over periods of 3–12 days, the maximum reduction in E_s compared to completely bare soil was 40%. This was observed under a maize canopy when the crop leaf area index (LAI) was 2.0. When the trees had similar values of LAI (1.9), the reduction in E_s was lower at 23%. However, in contrast over an entire rainy season, the presence of trees over bare soil reduced total seasonal E_s by 24%, compared to an 8% reduction in E_s by the maize crop. The maize crop was less effective in reducing E_s largely because of its shorter duration.

Qiu et al. [59] developed a method to estimate daily soil evaporation using differential measurements of temperature. They adopted the basic concept of differential temperature measurements to estimate soil evaporation. Evaporation was estimated from the temperature difference between wet soil and a reference dry soil in which negligible evaporation was assumed. Experiments were conducted in a field with sandy soil that was irrigated with sprinklers and included a weighing lysimeter to measure actual evaporation. Regression between modeled and actual evaporation on a daily basis produced a slope of 1.05 ($R^2 = 0.9$). It was observed that the cumulative evaporation calculated by the model was in reasonably good agreement with the actual value, both qualitatively and quantitatively.

Wythers et al. [76] measured soil water in lysimeters filled with three different soils. He measured daily evaporation gravimetrically and estimated evaporation rates using the energy balance method. The estimated soil water content at depths 3.8 cm, 11.4 cm, 19.0 cm, 26.6 cm, and 34.2 cm with time domain reflectometry rods and with soil cores. Bare-soil evaporation during the first 15 days of the experiment was 25% higher in the silt loam than the sandy loam, and 42% higher than the clay loam. By Day 30, bare-soil evaporation in all soils was 0.5 mm d^{-1}. Soil water content decreased in all five lysimeters layers, it was related to time and depth (r^2 values ranged from 0.60 to 0.95). The slope describing change in soil water content was greatest in the top 3.8-cm layer in all soil types (–0.52 in the clay loam, –0.32 in the silt loam, and –0.21 in the sand loam). At 14 day, bare-soil evaporation had the greatest influence on the upper 4.62 cm in the sand loam and the upper 7.18 cm in the clay and silt loams. At 51 day bare-soil evaporation had the greatest influence on the upper 7.36 cm in the sand loam, the upper 9.79 cm in the silt loam and the upper 14.1 cm in the clay loam.

2.2.2.3   Advanced Methods for Evaporation Measurements

Kumar et al. [43] estimated the daily soil water evaporation using an arti-ficial neural network. In field water balance studies, one of the major dif-ficulties was the separation of evapotranspiration into plant transpiration and soil evaporation components. In this study the *Radial Basis Function Neural Network* (RBFNN) was implemented using C language to estimate daily soil water evaporation from average relative air humidity, air tem-perature, wind speed and soil water content in a cactus field study. The RBFNN learned rapidly and converged after about 1000 training iterations. The optimum number of hidden neurons was found to be six. The RBFNN achieved good agreement between predicted and measured values. The average absolute error and the root mean squared errors were 21% and 0.17 mm for the RBFNN and 30.1% and 0.28 mm for the Multiple Linear Regression (MLR). The RBFNN technique appears to be an improvement over the MLR technique for estimating soil evaporation.

2.2.3   EVAPORATION FROM BARE SOIL

Lascano et al. [44] used a numerical model simulate evaporation, water and temperature profiles for a bare soil over drying cycles of 8, 9, and 20 days using soil hydraulic characteristics and meteorological data as input. The model was also used to evaluate the effect of the thickness of the surface layer in the model on the simulated evaporation rates. The simulated values were compared to actual measurements made over the same period of time. It was postulated that the model was highly accurate to predict the bare soil evaporation rates, as well as soil water and temperature profiles.

Matthias et al. [47] conducted two field studies to estimate bare soil evaporation for 7 days following surface trickle irrigation from a point-source emitter. A micro-lysimeter and an infrared thermometer were used to estimate evaporation at several sites from both wetted and non-wetted areas surrounding the emitter. Based on data collected from both methods it was concluded that evaporation accounted for about 33–40% of the applied water in both the studies. The micro-lysimeter method tended to estimate higher values than the infrared thermometer method. For dry soil conditions, the infrared thermometer method estimates were slightly higher. Agreement between the two methods was found to be good for the cumulative 7-day total evaporation estimates in both studies.

Soares et al. [70] used active microwave instruments to remotely sense the variations in soil moisture in space and time up to 10 cm below the surface. The thermal infrared radiance was also related to soil evaporation through the energy balance at the soil-air interface. If those two measurements were made simultaneously over a long period, it might be possible to monitor the actual evaporation and the soil water budget. Airborne remote sensing equipment gave the surface temperature and the surface water content of an essentially bare agricultural region (surface temperature from an infrared radiometer and the surface water content from radar). To validate the results, the sensible heat fluxes derived were compared with regional fluxes measured using a vertical Doppler SODAR.

Plauborg [55] studied the evaporation from bare soil in a temperate humid climate using micro-lysimeter and TDR. The use of TDR for measuring soil water content was investigated using a manual interpretation of the trace. It was reported that the accuracy of daily measurement was about +/− 0.5 mm H_2O day^{-1}. The micro-lysimeter method was not valid in periods with high precipitation. The use of time domain reflectometry (TDR) for measuring soil water content was investigated using a manual interpretation of the trace. The precision of changes in soil water content calculated from daily measurements with TDR was about 1.3 mm H_2O, when using probes of 50 cm length. However, improved precision may be obtained by the use of an automatic interpretation of the trace. Estimates of daily evaporation from bare soil calculated from the water balance equation and measurements of soil water content with TDR were compared with measurements with micro-lysimeters. The TDR technique was suitable for estimating bare soil evaporation when the soil water content was integrated over a 0–50 cm soil profile and drainage had ceased at the lower depths of the profile. Evaporation during a 13-day drying period in spring, just after the soil had been fully rewetted, was about 26 mm. During a 23-day drying period later in the season the evaporation from the bare soil was about 30 mm. In both periods the accumulated evaporation was rather high and equivalent to about 65% and 50% of the accumulated potential evapotranspiration in the first and second drying period, respectively, even though the soil water content in the 0–50 cm profile was well below field capacity at the beginning of the second drying period.

Qiu et al. [58] proposed a method for estimating evaporation from bare soil. The necessary input variables for the suggested method were temperature, net radiation and soil heat flux. There were three advantages of the proposed method over the conventional methods. In this method the soil surface resistance and aerodynamic resistance were not required. Secondly, the number of variables included in the method was fewer. Thirdly, measurement and analysis of the parameters involved are relatively easy. Sensitivity analysis showed that the suggested method is sensitive to temperatures.

Surface soil wetness determines whether evaporation occurs at the potential rate or is limited by soil water supply. Many land surface models calculate evaporation by parameterizing the relative humidity at the soil surface (a method) or the soil water diffusion resistance. The relationships of a and b to the average moisture content of soil surface layer thicknesses ranging from 0.5 to 10 cm were examined using Bowen ratio/energy balance measurements of evaporation from a bare loam/silt-loam soil at two adjacent sites, one of which was multi-packed while the other was disc-harrowed [75]. It was found that the relationships were sensitive to the surface layer thickness and tillage treatment and became better defined with larger thicknesses.

Denisov et al. [17] studied a method of calculation of soil moisture evaporation from bare soils, which was based on the theory of exchange processes in porous media. Studies were conducted to quantify the physical process of soil moisture evaporation. It was established that water vapor transfer in soil, as a porous medium, follows diffusion laws, and water vapor released into the atmosphere occurs due to turbulence. On the basis of this assumption, a theoretically grounded equation of soil moisture evaporation has been obtained. The parameters of the equation were the soil water content of soil and the characteristics of the surface atmospheric layer. Comparison of calculated evaporation data from bare soil, taking into account weather conditions for different climate zones and measured data, showed rather good agreement with evaporation values (correlation coefficient, $R = 0.71$).

Shriyangi and Kunio [67] developed an evaporation chamber for evaporation measurement. This device was unique in the sense that it used a chamber that was completely open at one end and thereby, minimizes the effect of the chamber on the natural profiles of temperature, humidity, and turbulence. It was used in estimating a newly formulated surface resistance to bare soil evaporation under dry topsoil conditions. Asimple energy balance model incorporating the depth of evaporating surface, blended with a new approach for describing surface resistance was developed and successfully tested with limited set of data obtained from a simple experiment, assuming ideal conditions. In addition to the newly formulated surface resistance, the depth of the dry soil layer was also estimated and was relatively comparable with measured value. The newly formulated surface resistance was found to be insignificant compared to the resistance imposed by the dry soil layer. The total surface resistance was modeled as a power function of soil moisture in the top 0–1 cm of soil, while the newly formulated resistance did not show any relation with the soil moisture.

Polyakov [56] developed an expression for calculating the evaporation intensity from bare wet soil derived from a joint consideration of heat and water flow dynamics in adjacent air and soil media. This expression refined well-known theoretical formulae to determine potential evaporation. Estimation was performed for the effect of transitional soil processes due to dramatic change in meteorological conditions and physical parameters characterizing soil state. In case of turf-podzolic soil the duration of the first stage, unsaturated soil water flows were computed at several depths of the water table and initial moisture distributions in the aeration zone. A preliminary estimation was made to study the effect of soil seal formed due to rainfall on the physical evaporation. It is established that soil compaction because of falling drops can noticeably intensify or reduce outflow from the soil surface. A theoretical analysis was also done to estimate the evaporation from the bare soil. Also, the evaporation in second and third stages were modeled based on a stationary model of consistent heat and water transfer in the system soil-atmosphere (subsurface layer). The effect of thermal, hydro-physical soil properties and meteorological elements on evaporation intensity and thickness of a dried layer was investigated. It was postulated that the soil hydraulic conductivity decides the movement of water. A boundary condition at the soil surface was found which reflected the peculiarities of water exchange between soil and air media at the stages under consideration. The calculations were performed for five widespread soil types.

Rushton et al. [65] introduced a new concept to account for continuing evapotranspiration on days following heavy rainfall even though a large soil moisture deficit exists. Algorithms for the computational model were provided. The methodology for recharge estimation using a soil moisture balance was applied to two contrasting case studies. The first case study referred to a rainfed crop in semi-arid northeast Nigeria, where ground water recharge occurs during the period of main crop growth. For the second case study in England, a location was selected where the long-term average rainfall and potential evapotranspiration were of similar magnitudes. For each case study, detailed information was presented about the selection of soil, crop and other parameters. The plausibility of the model outputs was examined using a variety of independent information and data. Uncertainties and variations in parameter values were explored using sensitivity analyzes. These two case studies indicate that the improved single-store soil moisture balance model is reliable approach for potential recharge estimation in a wide variety of situations.

Zhi et al. [78] carried out an experiment of measuring soil moisture by using dual-frequency microwave radiometer designed by the authors. The measured data were analyzed by using statistical regression method and the empirical regression models for retrieving soil moisture in L and C bands were developed. The soil moisture in a rainfall event was retrieved using the experiential regression model, which is consistent well with the field sampling value. The results showed that at low soil moisture (>75%) the brightness temperature is linear with soil moisture. However, when soil moisture is higher than 75%, the brightness temperature is not linear with soil moisture, so it is difficult for microwave radiometer to measure the changes of soil moisture.

Aydin et al. [7] studied that quantitative changes in evaporation from bare soils in the Mediterranean climate region of Turkey in response to the projections of a regional climate model (hereafter RCM) developed in Japan. Simulation of actual soil evaporation (Ea) was carried out using Aydin model combined with Aydin and Uygur model for estimating soil water potential of bare fields. It was found that the combination of models appeared to be useful in estimating water potential of soils and actual evaporation from bare soils, with only few parameters.

Braud et al. [13] studied the evaporation fluxes from bare soil columns, as well as to the corresponding isotopic composition of the water vapor, under non-steady state conditions. The experiment allowed an accurate determination of these quantities. The formulae propounded in the literature were used to estimate the isotopic composition of the evaporated water vapor. None of them was able to correctly reproduce the measured isotopic composition of water. The data were then used to estimate the value of the isotopic composition of the soil water, which should be used to get the right results for the isotopic composition of the evaporated water vapor. Results suggested that, when liquid transfer was dominant within the soil, the isotopic composition of evaporation could be controlled by the isotopic composition of the liquid water within very thin soil surface layers.

2.2.4 EVAPORATION STUDIES BASED ON ANALYSIS OF METROLOGICAL DATA

There are a number of meteorological parameters, which affect the rate of evaporation. The measurement by evaporation is mainly affected by variables such as vapor-pressure deficit, wind movement, soil temperature, air pressure, relative humidity, wind velocity; solar radiation and effective sun shine hours, etc. In this sub-section the review of some recent studies using the meteorological approach has been done.

Singh et al. [66] developed linear regression equations between weekly evaporation and different meteorological parameters to establish a suitable relationship between them on yearly and seasonal basis. Highest correlation was found to exist between evaporation and wind velocity in both seasonal and weekly analysis. Least correlation was found to exist between evaporation and sunshine hours. It was also noted that during summer season (April–June) correlation coefficient between evaporation and wind velocity was lower than in the rest of the year; due to the fact that during summer season the wind velocity was higher as compared to rest of the year when steady airflow conditions prevailed.

Hussain et al. [27] studied the relationship between observed and estimated crop evapotranspiration and its variation with other climatic parameters. The observed reference evapotranspiration was obtained by the ratio of measured crop evapotranspiration and crop coefficient published by FAO. Five methods to estimation of reference evapotranspiration were used. The modified Penman method made the best performance as it posses the real correlation with weather parameters.

Desborough et al. [18] collected atmospheric and land surface data from the HAPEX-MOBILHY field experiment those were used to compare the bare soil evaporation simulations of 13 land surface schemes and to examine the relationship between differences in evaporation and soil moisture. Computed total evaporation ranged between 100 and 250 mm for a 120-day period in which there were no vegetation present (bare soil). This large range in evaporation was not related to soil moisture differences. Prescribing surface soil moisture and temperature did not reduce the range in evaporation instead it increased. The model predictions of evaporation were in close agreement with each other when they were allowed to select their own surface conditions than when they were forced to use the same conditions.

Kimura et al. [40] developed a canopy model for estimating the diurnal and seasonal variations of heat balance component, surface temperature and soil water content. The model was composed of one canopy layer and two soil layers. The meteorological parameters used for the simulation were precipitation, temperature, sunshine duration, wind speed and vapor pressure.

Aydin et al. [7] studied quantitative changes in evaporation from bare soils in the Mediterranean climate region of Turkey in response to the projections of a regional climate model (RCM) developed in Japan. Daily RCM data for the estimation of reference evapotranspiration (ET_o) and soil evaporation (ET_s) were obtained for the periods of 1994–2003 and 2070–2079. Potential evaporation (PET) from bare soils was calculated using the Penman–Monteith equation with a surface resistance of zero. Simulation of actual soil evaporation (E_a) was carried out using Aydin model [7].

2.2.5 EVAPORATION MODELING

Kondo et al. [41] developed a simple model for evaporation from a bare soil surface. This model combined two processes of water vapor transport. The first process was the vapor transport in air expressed by the bulk formula, and the second, the molecular diffusion of vapor in the surface soil pores. The vapor was assumed to have been carried from the interior of the soil pores to the land surface. The resistance to the vapor diffusion in the soil pore was expressed using a new parameter, estimated by experimental means. General formulation of the so-called "surface moisture availability" was expressed with this model. The formulation showed that the "surface moisture availability" depended not only on volumetric soil moisture content but also on wind velocity and ratio of the specific humidity of the air. This was in good agreement with experimental findings with loam and sand under various conditions. A model was constructed for estimating evaporation from bare-soil surfaces. In the model, the evaporation was parameterized with the soil-water content for the upper 2 cm of the soil [42], and the heat and water transport within the soil layer below 2 cm was explicitly described by the heat conduction and moisture diffusion equations. Experiments on evaporation from loam packed in pans were also carried out. The present model simulated the observed evaporation and vertical profiles of soil temperature and water content very well.

Alvenas and Jansson [4] presented a modeling approach for predicting soil surface temperature and soil evaporation. The procedure was based on the equations for heat flow at the soil surface and included vapor diffusion and a semi-empirical correction function for the surface vapor pressure. The impact of changes in three important model parameters was studied by means of multiple model simulations. The first parameter determined the steepness of the water potential gradient close to the surface. The second parameter was the water vapor enhancement factor and the third one limits the lowest possible hydraulic conductivity during drying. Measurements of soil water content and soil temperature in a bare sandy loam were used to evaluate the models behavior. Results from the temperature tests indicated enhanced vapor diffusion and a probable value of the diffusion tortuosity coefficient close to 1.0, whereas a value close to 0.7 was more likely according to the soil water contents and calculated evaporation.

Bonachela et al. [12] formulated a functional model to estimate E_s at daily time steps. During the energy-limiting stage, E_s was calculated as the sum of the equilibrium evaporation at the soil surface and an aerodynamic term, derived from the Penman equation. For the falling rate stage, Ritchie's approach was adopted for the E_s calculations [62]. The model was successfully tested in an orchard of 6×6 m² spacing, typical of intensive olive orchards, under a wide range of evaporative demand conditions.

Snyder et al. [69] developed a model that used a daily mean evapotranspiration ET_o rate to estimate energy – limited soil evaporation (potential or Stage 1). It also used daily mean ET_o and a soil hydraulic factor β to estimate soil hydraulic property – limited evaporation (Stage 2). The model provided good estimates of cumulative soil evaporation on both hourly and daily basis when compared to observed soil evaporation in three field trials. Crop coefficient K_c values from cumulative hourly and cumulative daily soil evaporation estimates and ET_o data were compared. Using a soil hydraulic factor (β) equal to 2.6 the model gave a fair approximation for the widely used K_c curves for initial growth of crops (FAO-24). However, it was postulated that a site-specific β factor might improve soil evaporation and K_c estimates for site-specific applications.

Acs [1] analyzed the transpiration E_v and bare soil evaporation E_b processes, assuming homogeneous and inhomogeneous areal distributions of volumetric soil moisture content θ. For a homogeneous areal distribution of θ; a deterministic model was used, while for inhomogeneous distribution of θ; a statistical-deterministic diagnostic surface energy balance model was applied. In the experiments different parameterizations for vegetation and bare soil surface resistances and strong atmospheric forcing were used. The results suggested that Ev was much related non-linearly to environmental conditions than E_b. Both E_v and E_b depend on the distribution of θ, the wetness regime and the parameterization used. With the parameterizations, E_b showed greater variations than E_v. These results were valid for conditions having no advective effects or mesoscale circulation patterns and the unstable stratification.

Aydin et al. [6] tested a simple model for estimating evaporation from bare soils in different environments. The model was based on the relations among potential and actual soil evaporation and soil–water potential at the top surface layer of the soil, with some simplifying assumptions. Input parameters of the model were simple and relatively obtainable viz., climatic parameters for the calculations of potential soil evaporation and metric potential measured near the soil surface. And after experiments it was found that the simple daily time step model depending upon the period considered for the calculation of potential soil evaporation, was capable of estimating actual evaporation from soils as a function of matric potential measured at 1 cm soil depth accurately. But it was unable to provide good estimates in the case of using water potential at deeper layers.

Molina et al. [49] developed a multilayer model, based on the discretization of the pan water volume into several layers. The energy balance equations established at the water surface and within the successive in-depth layers were solved using an iterative numerical scheme. The model was calibrated and validated using hourly measurements of the evaporation rate and water temperature, collected in a Class-A pan located near Cartagena (Southeast Spain). The simulated outputs of both water temperature and E_{pan} proved to be realistic when compared to the observed values. Experimental data evidenced that the convective mixing process within the water volume induced a rapid homogenization of the temperature field within the whole water body. This result led them to propose a simplified version of the multilayer model, assuming an isothermal behavior of the pan.

Torres et al. [72] tested the FAO-56 model [2] on different conditions of evaporative demand and soil moisture. The data analysis showed that under low evaporative demand conditions the results from the model and the weighting lysimeter differ just about 7% for cumulative evaporation, while on high evaporative demand the results are more sensitive to initial soil humidity. The model assessed the bare soil evaporation more accurately when the initial water content was medium, while at low soil water content the model underestimated evaporation and at high soil water content evaporation was overestimated. These results allowed firmly concluded that the FAO-56 methodology might be implemented with confidence at regional scale in semi-arid conditions to estimate bare soil evaporation.

Malik et al. [46] developed a model for predicting evaporation from bare and freely drained soils with a deep water table. The model delineated three classical and one transitional drying stage as the drying front advances into the soil profile. Daily evaporation was estimated from the daily potential evaporation rate and depth of the drying front, attained at the start of the day. The approximating relations used for calculating soil evaporation rate as a function of advancing drying front under different drying stages. Input parameters of the model are simple and easily measurable under field conditions viz.; daily potential evaporation rate, wilting point and field capacity moisture contents.

Fatih [21] used the Penman equation that which calculated potential evaporation, to include in it the relative vapor pressure of an unsaturated soil to predict actual evaporation from a soil surface. This improved the prediction when the difference between the temperature of the soil surface and ambient air was relatively small. Although the new method provided accurate solutions for a wider range of water content from saturation to the lower limit of the liquid phase of a particular soil, the modification did not respond to the vapor phase of the soil moisture. Therefore, in the dry range (i.e., in the vapor phase in which the flow was entirely as vapor), either resistance models or a Fickian equation should be used.

Ranatunga et al. [60] provided an overview of soil water models and the basic modeling techniques employed by each model. Considerable emphasis was given to matching existing data availability with input data requirements for each model to identify the limitations of model application in terms of data availability. A comprehensive review of the application of soil water models was also given, supported by assessments of individual model performance. The limitations and assumptions made under various approaches to soil water modeling are subsequently examined.

2.3 MATERIALS AND METHODS

Bare soil evaporation is a complex and non-linear phenomenon because of its dependence on several interacting meteorological factors. Field experiments were conducted in Block No. 1 of the experimental farm of the Water Technology Centre, and Block No. 14 of IARI, New Delhi Farm to measure the bare soil evaporation from two different soil types. The laboratory simulation studies were also conducted in the Soil Physic Laboratory of the Water Technology Center, Indian Agricultural Research Institute, New Delhi. Mathematical models were also developed to establish the relationship between evaporation and meteorological variables. This section includes methodology adopted for field experimentation, laboratory simulation studies and model development.

2.3.1 *EXPERIMENTAL SITE*

The sites for conducting field experiments were located at the experimental farm of IARI, situated in West Delhi, India, between 28°37′22″ to 28°39′05″ N latitudes and 77°08′45″ to 77°10′24″ E longitudes at an average elevation of 230 m above the mean sea level.

Climate of Delhi is categorized as semi-arid, subtropical with hot dry summer and cold winter. It falls in the Agroecoregion – IV of India. The mean annual temperature is 25°C. The maximum temperature in the months of May–June reaches up to 45°C and an average minimum of 20°C is recorded during the month of December–January. The average annual rainfall is 714 mm, of which as much as 75% is received during the monsoon season (June–September). The average relative humidity varies from 34.1 to 97.9%. The average wind speed ranges between 0.45 to 3.96 m.s^{-1} [51].

2.3.1.1 Soil Types

Soils of IARI represent a typical alluvium profile of Yamuna origin. The entire IARI farm is covered with several soil series. The soil type ranges from sandy loam of Meharuli series to clay loam of the Jagat series [39].

The texture of the soil upto a depth of about 150 cm is almost uniform. As per USDA textural classification, major portion of the study area belongs to sandy loam class (SLC). There are only minor pockets representing clay, sandy clay and sandy clay loam textural classes. At some places, calcium layer with hard kankar exist below 150 cm depth. Porosity of the soils in general is about 40%. The soils of IARI farm belong to good class as far as its permeability is concerned.

To study the evaporation from different types of soils, two sites were selected having texturally different soils, namely: Sandy Loam (SL) and Silty Clay Loam (SCL). Two sites were specifically chosen due to clear distinction of soil properties. Soil samples were collected from different layers (from surface till depth of 0.6 m at an interval of 15 cm) and were analyzed to determine physical and chemical properties of SCL and SL.

2.3.1.1.1 *Chemical Properties of Soil*

The pH was determined by digital pH meter in 1:2 soil water suspensions. Electrical conductivity (EC) was determined by digital electrical conductivity meter in 1:2 soil water supernatant.

Experimental soil type-1 site was located in the Block No.1 of IARI, near WTC experimental plot. The chemical properties like pH and EC are presented in Table 2.1. The pH of the soil varied from 7.1 to 7.2 in different layers, which remained almost constant throughout the soil profile studied (Profile average pH was 7.18). The Electrical Conductivity varied from 0.11 to 0.17 dS/m. The average EC for the profile was 0.13 dS/m.

TABLE 2.1 Chemical Properties of the Soil of First Experimental Site

Depth (cm)	pH	EC (dS/m)	pH	EC (dS/m)
	Soil type-1 Sandy loam (SL)		Soil type-2 Silty clay loam (SCL)	
0–15	7.2	0.17	7.8	0.17
15–30	7.2	0.13	7.8	0.13
30–45	7.2	0.11	7.5	0.11
45–60	7.1	0.11	7.3	0.11
Average	7.18	0.13	7.6	0.13

Experimental soil type-2: The second field experiment was conducted on the Plot No. 2 of the Block No. 14 of experimental farm of IARI, New Delhi. The pH of the soil was 7.8, which reduced to 7.3 with the increase in soil depth up to 60 cm. The electrical conductivity varied from 0.17 to 0.11 dS/m from top to bottom layers (Table 2.1). The profile average of soil pH was 7.6 while the profile average of EC remained 0.13 dS/m, same as site 1.

2.3.1.1.2 Physical Properties of Soil

The soil physical properties were determined in the Soil Physics Laboratory of WTC, IARI, New Delhi, using the standard analytical procedures. The particle size distribution was analyzed using International Pipette Method [54]. Core cutters of standard dimensions were used for determination of soil bulk density. Soil moisture characteristic curve was determined using the Pressure Plate Apparatus. The soil physical properties are presented in Table 2.2.

Mechanical analysis of the soils was done by International Pipette Method [54]. The texture was determined from textural diagram of the International Society of Soil Science. The soil was saturated through capillary flow of water and the saturated moisture content was determined gravimetrically.

Soil moisture retention characteristics of both the experimental soils at various metric potentials were determined by Richard's Pressure Plate Apparatus [61]. The pressure chamber containing ceramic plates of 1, 5 and 15 bars was used. The soil moisture characteristic curves for two soils are shown in Figure 2.1.

TABLE 2.2   Physical Properties of Soil

Soil properties	Soil I	Soil II
Sand (%)	80.4	37.2
Silt (%)	4.9	28.1
Clay (%)	14.7	34.7
Texture	Sandy loam	Silty clay loam
Saturation percentage (by weight)	34.2	40.9

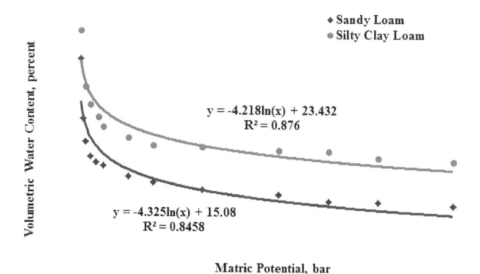

FIGURE 2.1   Relationship between soil matric potential and volumetric water content.

2.3.2   *EXPERIMENTAL PLOTS*

Experimental plots of 2×2 m^2 were selected for conducting field experiments. There were 3 plots in each field and a buffer strip of 1.5 m was maintained for separating the plots from one another. All vegetation, crop or grasses were removed, when necessary. The plots were kept bare throughout the experimental period. Pre-calibrated gypsum blocks were used to measure the soil water content in conjunction with tensiometers. Tensiometers of different sizes were installed in experimental plots to measure the soil water tension at soil depths of 15, 30, 45, and 60 cm. A corner of the plots was chosen for taking soil samples using gravimetric method. A screw auger was used for this purpose and the soil samples were transferred to the moisture canes soon after sampling. Care was taken to immediately transfer the samples to the laboratory for weighing and drying. The holes created during the gravimetric sampling were immediately filled with wet soil soon after taking the sample so that no cavities remained to affect the evaporation from deeper layers.

2.3.3 DETAILS OF LABORATORY SIMULATION STUDIES

2.3.3.1 Fabrication of Experimental Soil Columns

Rigid acrylic pipes, 310 mm in diameter × 600 mm long × 4 mm wall thickness, were used for containing soil columns of desired specification. All the pipes were sealed at the bottom by a square acrylic sheet of 8 mm thickness. The fabrication of experimental units was initiated by cutting the 8 mm acrylic sheet of the desired dimension and fixing by an adhesive to strongly join the edges of box. The rectangular sheet served not only as a physical support for the acrylic pipes filled with soil. The sheet was drilled with small holes of 1 mm diameter for an equal to the inner diameter (300 mm) of circular pipe. Holes of appropriate size were drilled at appropriate distances on bottom sheet purposefully to allow excess water to drain freely. The size of the hole and coverage area was designed based on the porosity of the soil. In addition to this, as a filter material, a porous cotton cloth was uses to avoid any mass wastage of soil during packing or saturation of the soil column. This was done to ensure that the small size soil particles are not flown down with soil.

2.3.3.2 Preparation of Soil Columns

Two texturally different soils [Soil I: Sandy loam (SL) and soil II: Silty clay loam (SCL)], were sampled from top 0–60 cm with 15 cm interval layer of Block No. 1 and Block No. 14 of Indian Agricultural Research Institute farm; respectively. The soils were air-dried ground to pass through 2 mm sieve before packing in the transparent acrylic soil columns. The soils were packed in the acrylic columns by varying the bulk density as measured at field up to 60 cm at 15 cm depth interval, in order to simulate the field conditions. The bulk densities of different layers of the experimental sites as measured in the field are given in Table 2.3.

For this purpose, weight of dry soil required to give the desired density in the space the acrylic pipes column was calculated. The soil was spread uniformly over a plain floor and covered with jute bags. Appropriate amount of distilled water was sprinkled uniformly over the bags and the

TABLE 2.3   Bulk Densities of Experimental Soils

Depth cm	Bulk density, gm cm⁻³	
	Sandy loam	**Silty clay loam**
0–15	1.51	1.43
15–30	1.56	1.45
30–45	1.58	1.50
45–60	1.57	1.53
Profile Average	1.56	1.48

soil was allowed to remain undisturbed for duration of 24 hours. Later on, the moist soil was thoroughly mixed to facilitate proper compaction while packing in the soil column. The moist soil thus prepared was compacted in the column in an incremental layer of 5 cm and so on so that the desired density could be attained. The soil surface was scratched before adding the successive increment of moist soil to insure proper contact between two soil layers and avoiding local compaction. The top 5 cm of the column was left unfilled to facilitate the saturation (Figure 2.2).

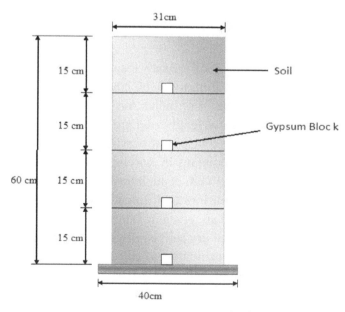

FIGURE 2.2   Schematic diagram of the soil column and soil packing.

To measure the bare soil evaporation from two different types of soil, a total of six numbers of soil columns (two treatments and three replications) were used that were especially fabricated in the workshop and Soil Physics Laboratory of W.T.C. All six-soil columns were placed in the safe enclosure to avoid any disturbances from external sources.

2.3.4 FABRICATION OF GYPSUM BLOCKS

Gypsum blocks of size 2 cm × 2 cm × 2 cm were specially fabricated, calibrated and installed at 15 cm depth intervals and at appropriate places in the field as well as soil columns to determine the *in-situ* soil moisture content (or corresponding suction) in different layers of soil columns by measuring their resistance using digital resistance meter (range 0–1000 ohms). A total number of 24 such gypsum blocks were fabricated by pouring gypsum slurry (gypsum:water; 1:1) into wooden molds in which two electrodes were already placed, separated by a distance of 0.5 cm and connected to copper wires for external connections. After the solidification of the slurry, the molds were opened, blocks taken out carefully and kept for sun drying for 5 days.

Gypsum blocks were calibrated for both types of soil. Half the number of blocks was placed in sandy loam and half in silty clay soils on two ceramic plates separately. Soil with embedded gypsum blocks on ceramic plates were saturated with water, each time for at least 24 hours, before subjecting these saturated soils on ceramic plates to a particular pressure in pressure plate apparatus. The pressure applied represented suction values of 0.1, 0.3, 0.5, 1, 2, 3, 5, 7, 10, 12, 15 bars. The resistance was measured for each block embedded in soil at different suctions. For each applied suction value, the gravimetric soil moisture content was also determined. Finally, a linear relationship was established between suction value and corresponding resistance value for each gypsum block. A digital multimeter was used for recording the electrical conductivity of the gypsum blocks in various layers in the soil columns.

2.3.5 TENSIOMETERS

Tensiometer consisted of water filled porous cup in contact with soil connected with water filled tube to a vacuum gauge. Water retained within the soil matrix came into equilibrium with the moisture in the porous cup, and since the system is closed, these forces were transmitted to the water column in contact with cup. When placed in dry soil water moved out of cup creating a negative suction that could be recorded using the manometer. In order to determine the soil moisture content at different depths in the experimental fields, the tensiometers and tube auger were used. The tensiometers of sizes 15, 30, 45, 60 cm were installed in the field at corresponding depths. Tensiometers provided direct measurement of soil moisture tension in the field and these readings were converted to the soil moisture using soil moisture characteristic curve of different soils.

2.3.6   TUBE AUGER

To determine moisture content at different depths, a tube auger was used in the field for collecting the soil samples from different soil depths.

2.3.7   CLIMATIC PARAMETERS

Keeping in view the objective of the research work and review of different models, the required data on climatic parameters were acquired from the meteorological observatories at the Division of Agricultural Physics and the Water Technology Center, IARI, New Delhi. The climate of Delhi region can be characterized as semi-arid, sub-tropical with hot dry summer and cold winter. Data pertaining to a total number of eleven climatic variables were collected for the period of study. However, while analyzing the data, it was observed that some data particularly related to solar radiation and wind speed were incorrect. Hence, these variables were not considered for model formulation although quite pertinent.

2.3.8   OBSERVATIONS

2.3.8.1   Soil Moisture Content

Soil moisture content was determined using gravimetric method. Soil samples were collected by tube auger sampler for different layers in 15 cm incremental depth up to 60 cm depth for the entire study period. Soil moisture samples were collected four times a day at four hours interval: starting from 8.00 a.m. in the morning till 8.00 p.m. in the night, to determine the depth of water evaporated in one day. The volumetric content of the soil moisture was determined from following equations:

$$\theta_w = \frac{Wetsoil - Drysoil}{Drysoil} \times 100 \tag{7}$$

$$\theta_v = \theta_w \times \rho_b \tag{8}$$

where, θ_w = soil moisture content on dry weight basis (%), θ_v = soil moisture content on volume basis (cm/m), and ρ_b = bulk density of the soil (g.cm^{-3}).

The field measurement of soil moisture contents in different soil layers were made using the procedure described as above. The bare soil evaporation was estimated on four hourly and daily bases. For this purpose, the final soil moisture of the layer was subtracted from the initial soil moisture. This way, the soil moistures of each layer were determined at four hours intervals and on daily bases. The experimentation started after fully saturation of the profiles in plots as well as column studies and after ensuring that the gravity drainage has completely ceased. It was assumed that all inflow and outflow were zero in the field study and care was taken to protect the soil columns from any moisture inflow particularly from rainfall for which appropriate measures were taken.

2.3.9 WATER BALANCE APPROACH FOR ESTIMATION OF BARE SOIL EVAPORATION

Rates of bare soil evaporation from experimental fields were measured using a simple water balance approach. For this purpose, the soil moisture contents of the profile at 15 cm interval up to a depth of 60 cm were measured using gravimetric and tensiometer methods. The loss of soil moisture was only attributed to evaporation, considering that there were no inflows or outflows from the boundary of experimental plots as well as columns; and no inflow such as rain or irrigation after starting the experiments.

2.3.10   MODEL DEVELOPMENT AND VALIDATION

Relationships between evaporation and climatic parameters were developed for top surface at the end of the study and validated with the known results after conducting the sensitivity analysis for the parameters. For this purpose, multiple linear regression (MLR) technique was applied to each set of data of two different soil types and regression analysis was carried out to obtain the values of different statistical parameters. The corresponding equations for prediction of evaporation from 15 cm soil depth for two soils were developed.

The following types of model equations were developed for bare soil evaporation by using appropriate standard statistical techniques of data fitting. The developed models were calibrated and validated using the field information collected in the present study.

$$\text{Bare soil evaporation } (E_s) = f \text{ (Meteorological parameters)} \qquad (9)$$

$$\text{Bare soil evaporation } (E_s) = f \text{ (maximum air temperature,}$$
$$\text{minimum air temperature,}$$
$$\text{maximum relative humidity,}$$
$$\text{minimum relative humidity,}$$
$$\text{solar radiation)} \qquad (10)$$

2.4   RESULTS AND DISCUSSION

The present study was under taken with objectives to quantify the bare soil evaporation from field and laboratory conditions. It was also indented to develop and validate models for prediction of bare soil evaporation and to determine the patterns of evaporation from deeper layers from sandy loam (SL) and silty clay loam (SCL). In this study, measurements of soil evaporation from soil columns and field plots for two different soil types and four depths were recorded over a long period of time. Results obtained from field experiments and laboratory simulations are presented in this section. Mathematical models were developed for bare soil evaporation relating the meteorological parameters. Model performances were also evaluated through statistical analysis: coefficient of determination (R^2)

and coefficient of correlation (r). The use of the developed models was made in predicting values of bare soil evaporation.

2.4.1   MEASUREMENT OF BARE SOIL EVAPORATION

2.4.1.1   Estimation of Bare Soil Evaporation in Experimental Field Conditions

It was found that the bare soil evaporation from all the layers under study varied from 1.8 to 3.9 mm per day in all the layers in SCL soil. In SL soil, the bare soil evaporation ranged from 1.3 to 3.7 mm per day during the period of studies. Upper two layers experienced high evaporation losses as compared to the deeper layers. The evaporation from 15 and 30 cm depths were approximately similar except in the initial few days after saturation. Whereas, the soil moisture contents in lower layers remained higher than upper layers indicating low evaporation. The general behavior of bare soil evaporation from the field experiments are shown in Figure 2.3.

It was observed that the evaporation from SCL soil ranged between 1–3.0 mm per day during 28th March, 2009 to 17th April, 2009. In the same period however, the evaporation losses from SL soil was found to be varying from 1.5 to 3.5 mm per day. It can be concluded therefore, that the evaporation from both soils during the reporting period did not vary too much. However, the layer wise comparison shows a clear-cut distinction between the bare soil evaporation of SCL and SL soils.

It was found that the bare soil evaporation from all the layers under study, varied from 1.0 to 3.9 mm per day in SCL soil. The upper layers experienced higher evaporation losses as compared to the deeper layers. The evaporation from 15 and 30 cm depths were approximately similar except in the initial few days after saturation. The soil moisture content in lower layers remained higher than upper layers indicating low evaporation. It was also recorded that the bare soil evaporation from the upper layers (0–15 and 15–30 cm) was constantly higher as compared to all other layers. The bare soil evaporation from 0–15 cm, 15–30 cm, 30–45 and 60 cm layers in SL soil varied between 2.0–2.8 mm per day after saturation (first phase), which increased to 3 mm per day in second phase (drying phase). The bare soil evaporation rose from 33 mm per day to 3.8 mm per day in

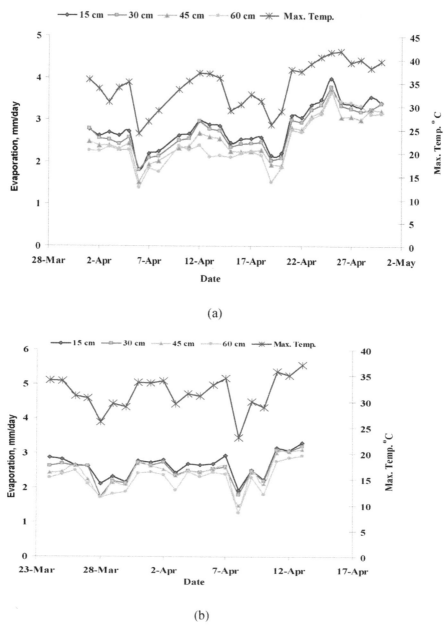

(a)

(b)

FIGURE 2.3   Bare soil evaporation from different depths. (a) Silty clay loam soil (field experiment); (b) sandy loam soil (field experiment).

the third phase when the root zone soil moisture reached wilting point. It was due to low water holding capacity of SL as compared to SCL soil. The pattern of change in evaporation was greatly affected by the ambient conditions, which had identical effect on both soils.

In both type of soils, the deeper layer (45–60) were not much affected by the ambient conditions. Therefore, this layer contained relatively higher soil moisture in the third phase when the rate of evaporation from lower most layers (60 cm) was recorded higher than all of the layers. It was due to the reason that while the soil moisture in all other layers have already been exhausted only this layer contained enough soil water for evaporation. It was observed that the patterns of evaporation in lower most layer were comparatively less similar in sandy loam soil as compared to silty clay loam soil because less amount of moisture was available in this type of soil for evaporation in later stages due to excess losses in previous phases.

2.4.1.1.1 Bare Soil Evaporation from Silty Clay Loam Soil in Experimental Field Conditions

The general behavior of bare soil evaporation from silty clay loam soil in the experimental field conditions is presented in Figure 2.3a. It was observed that the daily evaporation was almost constant up to five days after saturation in all layers, which decreased due to reduction in the ambient climatic conditions (reduction in temperature and increase in humidity) due to a minor amount of rainfall on 9th April, 2009. However, as the experiment field was well protected by covering the same with plastic sheet for not allowing any moisture to seep in to the ground, the reduction in the evaporation was mainly due to drop in temperature and increase in relative humidity.

It was observed that the bare soil evaporation from the upper layer (0–15 cm) was always higher as compared to all other layers. The bare soil evaporation from 0–15 cm, 15–30 cm, 30–45 cm and 60 cm layers varied between 1.8–2.8 mm per day after saturation (first phase), which rose to 3.0 mm per day in second phase (drying phase). The bare soil evaporation rose from 3.0 mm per day to 3.9 mm per day in the third phase when the root zone soil moisture reached wilting point. In the third phase when the rate of evaporation from lower most layer (60 cm) was recorded higher than all of the layers due to the reason that this layer contains relatively higher soil moisture while the soil moistures in all other layers have already been exhausted.

It is clear from Figure 2.3a that the evaporation rates decreased with increased the soil depth. The deepest layer has exhibited higher evaporation after 22nd April when the maximum temperature rose steadily from 25 to 40 degrees Celsius. With decrease in temperature from 43 to 38 degrees Celsius all layers exhibited a steady decline in the evaporation, but with varying rates.

2.4.1.1.2 *Bare Soil Evaporation from Sandy Loam Soil in Experimental Field Conditions*

It was found that the rate of evaporation in sandy loam soil (SL) was higher than silty clay loam soil (SCL). This is because the water holding capacity of clay particles are more due more porosity; hence, adhesive forces are higher. Evaporation rates measured at 15 cm depth intervals from sandy loam soil in field during the same duration are shown in Figure 2.3b.

It was observed that patterns of evaporation in lower most layer were comparatively less similar in sandy loam soil as compared to silty clay loam soil because less amount of moisture was available for evaporation in later stages due to excess losses in previous phases. In silty clay loam soil the differences in evaporation of different layers were not very high. In case of sandy loam soil there are large variations among the rates of evaporation in different layers as compared to silty clay loam soil Figure 2.3b.

2.4.1.2 Estimation of Bare Soil Evaporation in Experimental Laboratory Conditions (Column Study)

The evaporation from column studies for two different soils was measured by following the methodology as explained in the section on materials and methods. There were three replications for both types of soils. Evaporation from these replicates was averaged after measuring for drawing meaningful conclusions.

It was hypothesized that the evaporative area can have marked effect on bare soil evaporation from the soil columns as compared to the field experiments. It was however, not confirmed. The evaporation from columns varied from close to 1.0 mm per day to about 2.8 mm per day SCL soil during the periods of early April to early May. In the same period the evaporation from column studies in SL soil varied from 1.6 to 3.4-mm per day. The behavior of evaporation pattern of sandy loam soil was slightly different than silty clay soil. In sandy loam soil, three upper most layers (0–15, 15–30 and 30–45 cm) showed similar evaporation patterns, while the lower most layer (45–60 cm) showed lower rate of evaporation than upper layers (Figure 2.4b). There was marked reduction in evaporation rates due to the drop in temperature and higher humidity due to small amount of rain fall on 29 March and 9th April 2009 respectively that is reflected in Figure 2.4b.

The higher evaporation rates in the SL soil can be attributed to its low porosity due to which the water holding capacity was less. Initially the rates were high which reduced in later stages due to low availability of moisture for evaporation. Pattern of these variations can be attributed to the change in ambient climatic conditions.

In both type of soils the evaporation from the top surface layer (0–15 cm) was found to be higher than the bottom layers. Initially this system prevailed but due to the higher losses of moisture from 0–15 cm layer the moisture availability in this particular layer reduced. Hence, in the later stages, the evaporation from 15–30 cm, 30–45 cm and 45–60 cm layers have become more predominant due to more available moisture for evaporation. The deeper layers were less affected as compared to the top two layers due to the ambient climatic abrasions.

2.4.1.2.1 Estimation of Bare Soil Evaporation in Experimental Laboratory Conditions for Silty Clay Loam Soil

For silty clay loam soil, the evaporation from the column studies has been presented in Figure 2.4a. The perusal of the figure confirms the fact

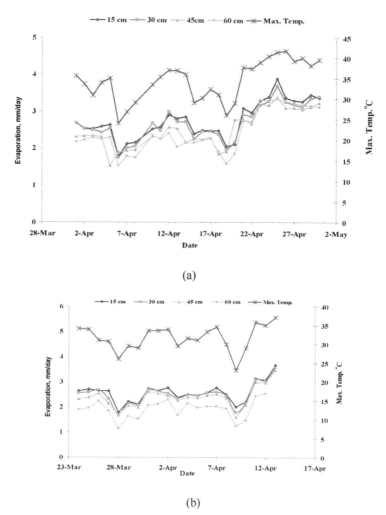

FIGURE 2.4   Evaporation from different depths (column study). (a) Silty clay loam soil (column study); (b) Sandy loam soil (column study).

that the evaporation from columns varied from close to 1.0 mm per day to about 3.8 mm per day during the periods of early April to early May. The variations were due to change in ambient climatic conditions.

The evaporation from the top surface layer (0–15 cm) was found to be higher than the bottom layers. Initially this system prevailed but due to the higher losses of moisture from 0–15 cm layer the moisture availability in this particular layer reduced. Hence, in the later stages the evaporation from 15–30 cm layer has become more due to more available moisture for evaporation. The deeper layers were less unaffected as compared to the top two layers. However, the change in rate of evaporation varied according to the variations in the temperature (Figure 2.4b).

2.4.1.2.2   Estimation of Bare Soil Evaporation in Experimental Laboratory Conditions for Sandy Loam Soil

The evaporation from column studies for sandy loam soil was also measured by following the methodology as explained in the section on materials and methods. After profile saturation it took less time for soil to come to field capacity moisture content. There were three replications for this soil too. In this section the results obtained from the laboratory simulation have been presented. The general pattern of evaporation was similar but the effect of changing ambient conditions was felt in deeper layers more in sandy soils than silty clay loam soil.

The behavior of evaporation pattern of sandy loam soil was slightly different than silty clay soil. In sandy loam soil, three upper most layers (0–15, 15–30 and 30–45 cm) showed similar evaporation patterns, while the lower most layer (45–60 cm) showed lower rate of evaporation than upper layers (Figure 2.4b). There was marked reduction in evaporation rates due to the drop in temperature and higher humidity due to small amount of rain fall on 29th March and 9th April 2009 respectively that is reflected in Figure 2.4b.

2.4.2   COMPARISON OF BARE SOIL EVAPORATION BEHAVIOR IN EXPERIMENTAL FIELD AND LABORATORY (SOIL COLUMN) STUDY

A comparison of field experiments and laboratory simulation was done to ascertain the effect of controlled conditions on evaporation behavior from two types of soils and development of mathematical models.

2.4.2.1   Comparison of Bare Soil Evaporation Behavior in Experimental Field and Laboratory (Soil Column) Studies for Silty Clay Loam Soil

A comparison of field experiments and laboratory simulation can be made. The evaporation rates obtained from field experiments were higher than the laboratory experiments throughout the period of study. The differences between the evaporations of 0–15 cm layers in experimental plots and laboratory simulation were not pronounced in SCL soil. In SL soil the evaporation rates obtained from field experiments were lower as compared to the laboratory experiments throughout the period of study except soon after rainfalls. The field and laboratory simulation responded identically to the changing ambient climatic conditions (Figure 2.5a). The difference between the evaporations of 0–15 cm layers in experimental plots and laboratory simulation was higher than silty clay loam soil but not much pronounced.

2.4.2.2   Comparison of Bare Soil Evaporation Behavior in Experimental Field and Laboratory (Soil Column) Studies for Sandy Loam Soil

In sandy loam soil, the evaporation rates obtained from field experiments were lower than the laboratory experiments throughout the period of study except soon after rainfalls. The field and laboratory simulation responded identically to the changing ambient climatic conditions in this case too (Figure 2.5b). The difference between the evaporations of 0–15 cm layers in experimental plots and laboratory simulation was higher than silty clay loam soil but not pronounced.

2.4.3   EVAPORATION IN RELATION TO METEOROLOGICAL PARAMETERS

Evaporation from bare soil surface depends not only on soil properties but also on atmospheric conditions. Evaporation is mainly affected by the supply of heat and the vapor pressure gradient, which in turn depend on

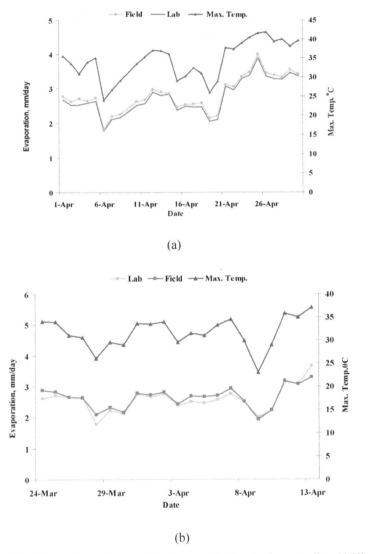

FIGURE 2.5   Comparison of evaporation between field and column studies. (a) Silty clay loam soil; (b) Sandy loam soil.

different meteorological parameters. The significance of air temperature, relative humidity, bright sunshine duration and wind speed influencing evaporation and soil type influencing evaporation was evaluated by linear, multiple regression and correlation analysis.

2.4.3.1   Correlation Analysis Between Bare Soil Evaporation and Meteorological Parameters for Sandy Clay Loam (SCL) Soil

The correlation coefficient between bare soil evaporation from SCL soil and meteorological parameters such as maximum temperature, minimum temperature, RH1, RH2 and solar radiation were estimated to be 0.95, 0.96, −0.94, and −0.95, respectively. It was felt necessary to fit the linear regression but a loss of R^2 was observed. It was concluded that there was a positive correlation between bare soil evaporation and meteorological parameters. However, the relative importance of the meteorological parameter was ranked based on correlation coefficient (r) and it was found that for SCL soil the meteorological parameters affecting the bare soil evaporation can be arranged in sequence: maximum temperature, minimum temperature, RH1, RH2 and solar radiation. The wind speed and solar radiation data being erroneous were left out of the analysis, but it is a known fact that this has very strong positive correlation with evaporation. The relations among the evaporation from SCL soil and maximum temperature, maximum relative humidity and minimum relative humidity are shown in Figure 2.6.

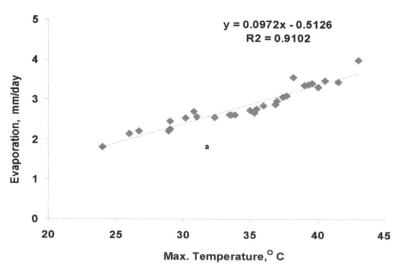

FIGURE 2.6A   Relationship between daily-observed evaporation and daily maximum temperature: sandy clay loam (SCL) soil.

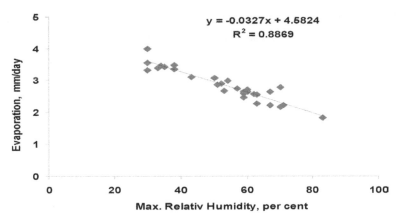

FIGURE 2.6B Relationship between daily evaporation and maximum relative humidity: sandy clay loam (SCL) soil.

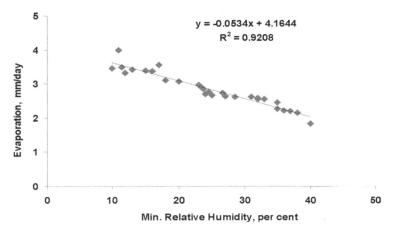

FIGURE 2.6C Relationship between daily evaporation and minimum relative humidity: sandy clay loam (SCL) soil.

2.4.3.2 Correlation Analysis Between Bare Soil Evaporation and Meteorological Parameters for Sandy Loam Soil

The correlation coefficients between bare soil from SL soil and maximum temperature, minimum temperature, RH1, and RH2 were 0.96, 0.94, −0.91, and −0.79, respectively. It was concluded that there was good correlation between bare soil evaporation and meteorological parameters. However; the relative importance of the meteorological parameter was ranked based on correlation coefficient (r). It was also found that for SL soil the meteorological parameter affecting the bare soil evaporation can be arranged in sequence: maximum temperature, minimum temperature, RH1, RH2. For SL soils, these linear regressions are shown in Figure 2.7.

2.4.3.3   Linear Regression Analysis of Evaporation for Silty Clay Loam Soil

Regression equations of polynomial type of third order were also fitted between meteorological parameters and evaporation. Fitting the third order polynomial functions was not too difficult using the computerized pro-grams. The predictions by interpolations as well as extrapolations, do suffer due to the fact that the polynomials are known to change shapes. Therefore, it was found appropriate that linear relations should also be tried. Finally the linear relations have been retained due to the fact that the bare soil evaporation was known to vary linearly with meteorological parameters.

Maximum air temperature has been found to correlate well with soil evaporation at both the experimental sites. Maximum and minimum rela-tive humidity have shown consistently the negative correlation with soil evaporation at both sites. This confirms that evaporation is directly related with air temperature and indirectly with relative humidity. The higher the temperature, warmer is air and the rate of evaporation is also high. Due to

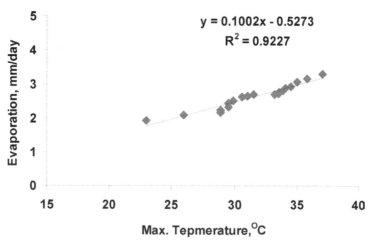

FIGURE 2.7A   Relationship between daily-observed evaporation and daily maximum temperature: sandy loam soil.

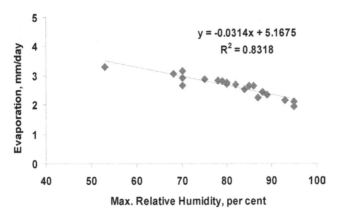

FIGURE 2.7B Relationship between daily-observed evaporation and daily maximum relative humidity: sandy loam soil.

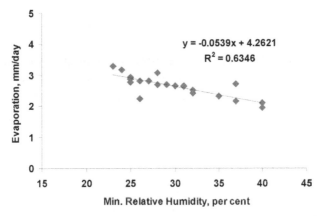

FIGURE 2.7C Relationship between daily-observed evaporation and daily minimum relative humidity: sandy loam soil.

this reason, the evaporation loss is comparatively more in summer season than in winter season. The regression results are highly significant. Pawar [52], Sharma and Hassan [26, 66] had observed a significant correlation between evaporation and maximum temperature (T_{max}).

2.4.3.4 Linear Regression Analysis of Bare Soil Evaporation from Sandy Loam Soil

There is a positive correlation between daily maximum temperature and daily-observed evaporation from bare soil. Pawar [52] and Hassan

[26, 66] also reported that relative humidity (RH) had an inverse correlation with evaporation. The higher the atmospheric humidity, the lower will be the evaporation. During evaporation process, the water molecule move from higher moisture to lower moisture levels. The rate of this movement was governed by difference of moisture content or moisture gradient of the atmospheric air. Hence, if humidity is more, the evaporation will be less and vice versa. In this study, it was recorded that the evaporation reduced with reduction in temperature and increase in relative humidity.

2.4.4 DEVELOPMENT OF EVAPORATION ESTIMATION EQUATIONS (EVAPORATION MODELING)

Evaporation estimation equations on daily basis were developed from the combined meteorological parameters using multiple linear regression (MLR) analysis. Multiple linear regression technique was purposefully applied to each set of data as the program for data analysis was available for MLR. However, nonlinear multiple regression (MNLR) technique could have served our purpose better but the availability of the appropriate software restricted the analysis. Nevertheless, statistical analysis carried out with the data for month from March to April on daily basis showed that the values of regression coefficients were positive for temperature, sunshine hour and wind velocity indicating that increment in any of these values would enhance the rate of evaporation. The developed models for SCL and SL soils are shown below:

For SCL soil:
$$Es = C + F\ (B, X) = C + B1\ X1 + B2\ X2 + B3\ X3 +$$
$$B4\ X4 = 2.59 + 0.0380\ (T_{max}) - 0.0021\ (T_{min}) -$$
$$0.0130(RH1) - 0.138\ (RH2),\ R^2 = 0.95 \qquad (11)$$

For SL soil:
$$Es = C + F\ (B, X) = C + B1\ X1 + B2X2 + B3\ X3 +$$
$$B4\ X4 = 2.15 + 0.0632\ (T_{max}) - 0.0090\ (T_{min}) -$$
$$0.0176\ (RH1) - 0.0006\ (RH2),\ R^2 = 0.87 \qquad (12)$$

2.4.5 MODEL VALIDATION

To see the combined effect of all parameters on evaporation, initially multiple linear regression models were developed using the meteorological parameters (maximum and minimum temperature, maximum and minimum relative humidity, sunshine duration and wind speed). However, on finding that the solar radiation and wind speed data were erroneous they were left out from the model development. Model development for all the layers were attempted but it was found appropriate that only 0–15 cm layer is affected more due to the changing ambient conditions. Hence, models for 0–15 cm layers only were considered for validation. The model development was done based on the laboratory experiments and validation with field experimental data for both soil types.

The parameters were incorporated into single equation and analyzed for t-statistic test for estimation of significance among the meteorological parameters. Meteorological parameters with non-significant t value for their coefficients were removed from the estimation equation. Since the evaporation takes place at topsoil surface and atmospheric interphase, the models were developed for 0–15 cm soil layers only.

The other details for statistics analysis are given in Tables 2.4 and 2.5. Daily evaporations were estimated from the developed multiple linear regression models. The estimated values were then compared with the observed values of evaporation as shown in Figure 2.8 for both soils. The developed equations were validated with observed information and relations between observed evaporation and predicted evaporation for silty clay soil and sandy loam soil were shown in Figure 2.8. After validation, it was found that both equations predicted values close to observed values. Hence, developed equation to predict the bare soil evaporation showed high value of accuracy.

Maximum temperature, wind velocity and solar radiation are highly positively correlated with the bare soil evaporation. The correlation coefficient between bare soil evaporation for SCL soil and meteorological parameters (maximum temperature, minimum temperature, RH1, and RH2) were 0.95, 0.96, −0.94, and −0.95, respectively. The correlation coefficient between bare soil evaporation for SL soil and meteorological parameters were 0.96, 0.94, −0.91 and −0.79, respectively. Pawar [52],

TABLE 2.4   Meteorological Parameters Influencing the Evaporation and Their Statistical Significance for SCL Soil (15 cm)

Predictor variable (Xi)	Regression coefficient, (Bi)	Standard error	t-value
Max. Temperature (T), °C	0.0380	0.897	2.239
Min. Temperature (T), °C	−0.0021	0.017	−0.261
Relative humidity (RH1), %	−0.0130	0.0080	−2.945
Relative humidity (RH2), %	−0.0138	0.0044	−1.284
Constant (C)	2.59	0.010	2.288

Multiple, R = 0.097

Coefficient of determination, R^2 = 0.95

Standard error = 0.120

Adjusted, R = 0.94

Prediction equation for, bare soil evaporation, from SCL soil (15 cm):

$Es = C+F (B, X) = C+ B1\ X1+ B2X2 +B3\ X3 + B4\ X4 = 2.59 + 0.0380\ (T_{max}) + 0.0021\ (T_{min}) - 0.0130\ (RH1) - 0.0138\ (RH2)$

TABLE 2.5   Meteorological Parameters Influencing the Evaporation and Their Statistical Significance for SL Soil (15 cm)

Predictor variable (Xi)	Regression coefficient (Bi)	Standard error	t-value
Max. Temperature (T), °C	0.0632	0.68	5.57
Min. Temperature (T), °C	−0.0090	0.024	−0.05
Relative humidity (RH1), %	−0.0176	0.0085	−3.94
Relative humidity (RH2), %	−0.0006	0.0073	−2.39
Constant, (C)	2.1	1.331	1.61

Multiple R = 0.98

Coefficient of determination, R^2 = 0.87

Standard error = 0.15

Adjusted, R = 0.93

Prediction equation for, bare soil evaporation, from SL soil (15 cm)

$Es = C+ F(B, X) = C+ B1\ X1+ B2X2 + B3\ X3 + B4\ X4 = 2.15 + 0.0632\ (T_{max}) - 0.0090\ (T_{min}) - 0.0176\ (RH1) - 0.0006\ (RH2)$

Sharma and Hassan [26, 66] had observed a significant correlation between evaporation and maximum temperature (T_{max}). Consequently with increased temperatures, the bare soil evaporation has shown an

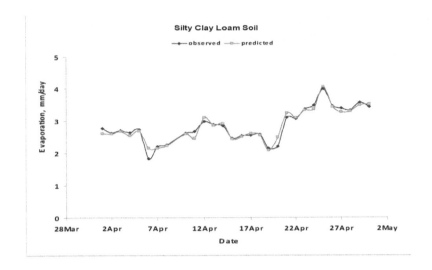

a

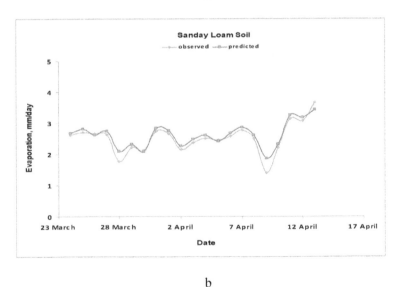

b

FIGURE 2.8 Comparison between observed evaporation and predicted evaporation. (a) Silty clay loam (SCL) soil; (b) Sandy loam (SL) soil.

increasing trend, whereas with lowering in temperature due to rainfall, the bare soil evaporation reduced in both the soils. Humidity was negatively correlated with bare soil evaporation, which has been reflected in the trends of evaporation from both soil types. These effects are not felt by deeper layers in the same intensity as the top layers and that has been reflected in the figures.

Many approaches have been tried in past to model the bare soil evaporation. Some of them were too simple and some of them too complex. A detailed review on this aspect has been presented in this chapter. At present, looking to the limitations of instrumentations, time availability and availability of meteorological parameters, it was considered to attempt the fitting of multiple linear regression equations for modeling bare soil evaporation knowing fully well that evaporation is a complex non-linear phenomenon. The main emphasis, however, was on simplifying the process for its wider applicability and ease of operation by less technical manpower mainly farming community for its possible field application in and around Delhi.

Model validation resulted in close agreement of observed and predicted evaporations for both type of soils. It is already known that many complex models already exist in literature, which requires collection of large data for their possible application. In this regard, it suffices here to mention that the present model can be used for predicting bare soil evaporation from SCL soil and SL soil with fairly good accuracy. Also as the present model requires less inputs, it can be used by the users in far-flung areas of Delhi where standard Class-A pan observatories are not available, but a few instruments to measure maximum and minimum temperatures and relative humidity are on hand.

2.5   SUMMARY

Field and laboratory experiments were conducted to measure the bare soil evaporation from silty clay loam soil of Jagat series and sandy loam soil of Mehrauli series of IARI farm, New Delhi during the months of March–April 2009. Soil moisture content in different layers (0–15, 15–30, 30–45 and 45–60 cm depths) below ground level (b.g.l.) were measured using gypsum blocks, tensiometers and gravimetric methods at a 4 hours interval daily after field saturation and ensuring the absence of any influx or efflux. To measure the evaporation from soil columns gypsum blocks of size 2 cm × 2 cm × 2 cm were specially prepared and used to determine the *in-situ* soil moisture content in different layers of soil columns by measuring their resistance. Bare soil evaporation was estimated following a water balance approach from the measured soil moisture content. Laboratory simulation studies were also conducted to characterize the behavior of different soil types. Quantitative estimates of bare soil evaporation from identical depths, as in case of field experiments, were made during the same period as field experiments.

In order to develop bare soil evaporation prediction models, the evaporation patterns from soil columns were considered for 0–15 cm soil depth only. It was assumed that the effects of meteorological variables are more profound in the top surface as compared to deeper layers. Linear multiple regression modeling technique was used to develop the models for predicting bare soil evaporation using air temperature, relative humidity and solar radiation.

The results obtained from the present studies indicated the bare soil evaporation from 0–15 cm soil layer (top layer) in silty clay loam soil varied from 2.0 to 3.0 mm per day in the first phase which rose to 3.9 mm per day in the drying stage. For sandy loam soil, the bare soil evaporation varied from a minimum of 2.0 mm to a maximum of 3.8 mm per day in uppermost layer. In both soils, the changes in soil moisture in other two layers were large though, the same cannot be referred to as evaporation in strict sense. The laboratory simulation with two soils indicated almost identical behavior. The bare soil evaporation from the laboratory simulation was recorded less as compared to field condition due to the controlled environmental conditions.

The model validation was undertaken with the data collected from the field studies. The model validation with observed bare soil evaporation data from field resulted in a close agreement with predicted. Coefficient of determination (R^2) for silty clay loam soil was 0.95 and for sandy loam soil was 0.87. The study concluded that the bare soil evaporation can be predicted reasonably well by the equations (developed models) for both soils for 0–15 cm soil depth. Since the agro-climatology of the Delhi region does not vary much, these results can be successfully applied to this region outside IARI for the assessment of bare soil evaporation from similar soil types. From this study the following major conclusions can be drawn:

1. The ambient temperatures showed maximum influence on bare soil evaporation in both soil types.
2. Relative humidity is negatively correlated with bare soil evaporation.
3. Multiple linear regression models can be used fairly accurately for estimation of evaporation from SCL and SL soils of IARI farm.
4. There were marked differences in the evaporation behavior of SCL and SL soils, wherein the evaporation rates were higher in SCL as compared to SL soils.
5. During the Months of March and April 2009, the evaporation rates in SCL soils varied from 1.8 to 3.9 mm per day and in SL soil 2.0 to 3.8 mm per day.

2.6   SUGGESTIONS FOR FUTURE WORK

Based on the results of the present study and discussion thereafter, it was felt that more attention should be paid to the following points for future research work to get better results:

a. Sensitivity analysis of the meteorological parameters should be done prior to developing modeling strategies.
b. To improve the performance of regression model, measurements of the soil temperature should also be made.
c. Use of improved instruments to measure the soil evaporation and a data logger facility would be more appropriate.

KEYWORDS

- bare soil
- crop water requirement
- evaporation
- evapotranspiration
- FAO
- IARI

- **India**
- **irrigation**
- **multiple regression analysis**
- **relative humidity**
- **transpiration**
- **water balance method**

REFERENCES

1. Acs, F. (2003). A comparative analysis of transpiration and bare soil evaporation. *Boundary-Layer Meteorology*, 109, 139–162.
2. Allen, R. G., Pereira, L. S., Raes, D., & Smith, M. (1998). *Crop Evapotranspiration: Guidelines for Computing Crop Water Requirements*. FAO Irrigation and Drainage Paper-56, Food and Agricultural Organization of the United Nations (FAO), Rome, Italy, p. 78–86.
3. Allen, R. G., Pereira, L. S., Smith, M., Raes, D., & Wright, J. L. (2005). FAO-56 dual crop coefficient procedure for predicting evaporation from soil and application extensions. *J. Irrig. Drain. Eng.*, 131(1), 65–71.
4. Alvenas, G., & Jansson, P. E. (1997). Model for evaporation, moisture and temperature of bare soil: calibration and sensitivity analysis. *Agric. For. Meteorol.*, 88, 47–56.
5. Aydin, M. (1994). Hydraulic properties and water balance of a clay soil cropped with cotton. *Irrig. Sci.*, 15, 17–23.
6. Aydin, M., Yangb, S. L., Kurt, N., & Yanoc, T. (2005). Test of a simple model for estimating evaporation from bare soils in different environments. *J. Ecol. Model.*, 182, 91–105.
7. Aydin, M., Yano, T., Evrendilek, F., & Uygur, V. (2008). Implications of climate change for evaporation from bare soils in a Mediterranean environment. *Environ. Monit. Assess.*, 140, 123–130.
8. Beese, F., Ploeg, R. R., & Richter, W. (1977). Test of a soil water model under field conditions. *Soil Sci. Soc. Am. J.*, 41, 979–984.
9. Black, T. A., Gardner, H. R., & Thurtell, G. W. (1969). The predictions of evaporation, drainage and soil water storage for a bare soil. *Soil Sci. Soc. Am. J.*, 33, 655–660.
10. Boast, C. W., & Robertson, T. M. (1982). A "micro-lysimeter" method for determining evaporation from bare soil: description and laboratory evaluation. *Soil Sci. Soc. Am. J.*, 46, 689–696.
11. Boesten, J. J. T. I., & Stroosnijder, L. (1986). Simple model for daily evaporation from fallow tilled soil under spring conditions in a temperate climate. *Netherlands J. Agric. Sci.*, 34, 75–90.

12. Bonachela, S., Orgaz, F., Villalobos, F. J., & Fereres, E. (1999). Measurement and simulation of evaporation from soil in olive orchards. *J. Irrig Sci.,* 18, 205–211.
13. Braud, I., Biron, P., Bariac, T., Richard, P., Canale, L., Gaudet, J. P., & Vauclin, M. (2009). Isotopic composition of bare soil evaporated water vapor. Part I: RUBIC- IV experimental setup and results. *J. Hydrol.,* 369, 1–16.
14. Brisson, N., Itier, B., Hotel, J. C., & Lorendeau, J. Y. (1998). Parameterization of Shuttleworth Wallace model to estimate daily maximum transpiration for use in crop models. *J. Ecol. Model.,* 107, 159–169.
15. Christopher, B. G. (2005). Evaporation from bare soil and soil cracks: A numeric study. Unpub. M. Sc. Thesis, The Oregon State University, USA.
16. Deki, L. J., Mihailovi, D. T., & Rajkovi, B. (2003). A study of the sensitivity of bare soil evaporation schemes to soil surface wetness, using the coupled soil moisture and surface temperature prediction model. *Meteorology and Atmospheric Physics.,* 23, 101–112.
17. Denisov, Y. M., Sergeev, A. I., Bezborodov, G. A., & Bezborodov, Y. G. (2002). Moisture evaporation from bare soils. *J. Irrig. Drain. Eng.,* 16, 175–182.
18. Desborough, C. E., Pitman, A. J., & Irannejad, P. (1996). Analysis of the relationship between bare soil evaporation and soil moisture simulated by 13 land surface schemes for a simple non-vegetated site. *Global and Planetary Change,* 13, 47–56.
19. Droogers, P., & Bastiaanssen, W. (2002). Irrigation performance using hydrological and remote sensing modeling. *J. Irrig. Drain. Eng.,* 128 (1), 11–18.
20. Ehlers, W., & Ploeg, R. R. (1976). Evaporation, drainage and unsaturated hydraulic conductivity of tilled and untilled fallow soil. *Zeitschrift für Pflanzenernährungund Bodenkunde,* 139, 373–386.
21. Fatih, K. (2007). Modification of the Penman method for computing bare soil evaporation. *Hydrological Processes,* 21, 3627–3634.
22. Garratt, J. (1993). Sensitivity of climate simulations to land-surface and atmospheric boundary – layer treatments – A review. *J. Climate,* 6, 419–449.
23. Ghildyal, B. P., & Tripathy, R. S. (1987). *Soil Physics.* Wiley Eastern Publishers, New Delhi. 656pp.
24. Gill, K. S., & Prihar, S. S. (1983). Cultivation and evaporativity effects on the drying patterns of sandy loam soil. *J Soil Sci.,* 135, 367–76.
25. Hanks, R. J., & Hill, R. W. (1980). *MODELING Crop Response to Irrigation in relation to Soils, Climate and Salinity.* International Irrigation Information Center Publi. No. 6, Pergamon Press, Elmsford, NY, USA.
26. Hassan, B. (1998). Comparative hydrometeorology of temperate and semiarid environments in India. *Drought Network News,* 10(3), 7–9.
27. Hillel, D. (1980). *Applications of Soil Physics.* Academic Press, New York, USA.
28. Hussain, Q., Mushabbir, P. M., & Christoph, F. (1990). Relationship between observed and estimated crop evapotranspiration and climate variations. *Agricultural Mechanization in Asia, Africa and Latin America.* 21(3), 61–65.
29. IPCC, (2001). *Climate Change 2001: The Scientific Basis.* Cambridge University Press. Cambridge.
30. Jackson, N. A., & Wallace, J. S. (1999). Soil evaporation measurements in an agroforestry system in Kenya. *Agric. For. Meteorol.,* 94 (3–4), 203–215.

31. Jackson, R. D., Idso, S. B., & Reginato, R. J. (1976). Calculation of evaporation rates during transition from energy-limiting to soil-limiting phases using albedo data. *Water Resour. Res.,* 12, 23–26.

32. Jackson, T. J., Levine, D. M., & Griffis, A. (1993). Soil moisture and rainfall estimation over a semiarid environment with the ESTAR microwave radiometer. *Transactions on Geoscience and Remote Sensing*, 31, 836–841.

33. Jacobs, A. F. G., & De Bruin, H. A. R. (1998). Makkink's equation for evapotranspiration applied to unstressed maize. *Hydrological Processes*, 12, 1063–1066.

34. Jalota, S. K., & Prihar, S. S. (1998). *Reducing Soil–Water Evaporation with Tillage and Straw Mulching*. Iowa State University Press, Ames, 142 pp.

35. Jalota, S. K., & Prihar, S. S. (1987). Observed and predicted trends from a sandy loam soil under constant and staggered evaporativity. *Aust. J. Soil Res.,* 25, 243–249.

36. Jalota, S. K., & Prihar, S. S. (1986). Effects of atmospheric evaporativity, soil type and redistribution time on evaporation from bare soil. *Aust. J. Soil Res.,* 24, 357–366.

37. Johns, G. G. (1982). Measurement and simulation of evaporation from a red earth. I. Measurement in a glasshouse using a neutron moisture meter. Aust. *J. Soil Res.,* 20, 165–178.

38. Kanemasu, E. T., Stone, L. R., & Powers, W. L. (1976). Evapotranspiration model tested for soybean and sorghum. *Agron. J.,* 68, 569–572.

39. Khalil, A., Singh, D. K, Singh, A. K., & Khanna, M. (2007). Modeling of nitrogen leaching from experimental onion field under drip fertigation. *J. Agric. Water Manage.,* 89, 15–28.

40. Kimura, R., Kondo, J., Otsuki, K., & Kamichika, M. (1999). A prediction model for evapotranspiration from vegetative surface model and experimental verification. *J. Jpn. Soc. Hydrol. Wat. Res.,* 12(1), 17–27.

41. Kondo, J., Saigusa, N., & Sato, T. (1990). A parameterization of evaporation from bare soil surfaces, *J. Appl. Meteorol.,* 29 (5), 385–389.

42. Kondo, J., Saigusa, N., & Sato, T. (1992). A model and experimental study of evaporation from bare-soil surfaces. *J. Appl. Meteorol.,* 31, 304–312.

43. Kumar, M., Raghuwanshi, N. S., Singh, W. R., Wallender, W., & Pruitt, W. O. (2002). Estimating Evapotranspiration using Artificial Neural Network. *J. Irrig. Drain. Eng.,* 128(4), 224–233.

44. Lascano, R. J., & Baval, V. (1986). Simulation and measurement of evaporation from a bare soil. *Soil Sci. Soc. Am. J.,* 50, 1128–1133.

45. Linsley, R. K., Kohler, M. A., & Paulhus, J. L. H. (1992). Evaporation and Transpiration. In: *Hydrology for Engineers*. McGraw Hill, New York.

46. Malik, R. K., Anlauf, R., & Richter, J. (2007). A simple model for predicting evaporation from bare soils. *J. Earth and Environmental Science*, 155(4), 293–299.

47. Matthias, A. D., Salehi, R., & Warrick, A. W. (1986). Bare soil evaporation near a surface point-source emitter. *Agric. Water Manage.,* 11(3–4), 257–267.

48. Mellouli, H. J., Wesemael, B., Poesen, J., & Hartmann, R. (2000). Evaporation losses from bare soils as influenced by cultivation techniques in semi-arid regions. *Agric. Water Manage.,* 42, 355–369.

49. Molina, J. M., Alvarez, V. M., Gonza, M. M., & Baille, A. (2006). A simulation model for predicting hourly pan evaporation from meteorological data. *J. Hydrol.,* 318, 250–261.

50. Monteith, J. L., & Unsworth, M. H. (1990). *Principles of Environmental Physics.* 2nd Edn. Edward Arnold, London.
51. Nathan, K. K. (2007). *Annual Report of IARI 2006–2007.* New Delhi, India.
52. Pawar, H. L. (1993). Relationship between evaporation USWB class A pan evaporimeter and metrological parameters. *J. Water Manage.,* 1(2), 103–104.
53. Peck, A. J., Hatton, T. (2003). Salinity and the discharge of salts from catchments in Australia. *J. Hydrol., 272,* 191–202.
54. Piper, S. C. (1966). *Soil and Plant Analysis: A laboratory manual of methods for the examination of soils and the determination of the inorganic constituents of plants.* Interscience Publishers, Inc., New, York, 384 pp.
55. Plauborg, F. (1995). Evaporation from bare soil in a temperate humid climate – measurement using micro-lysimeters and time domain reflectometry *Agric. For. Meteorol.,* 76(1), 1–17.
56. Polyakov, V. L. (2005). Simulation of evaporation from bare soil without and with the soil surface seal. *Intl. J. Fluid Mech. Res.,* 32(2), 214–254.
57. Priestley, C. H. B., & Taylor, R. J. (1972). On the assessment on surface heat flux and evaporation using large-scale parameters. *Monthly Weather Rev.,* 100, 81–92.
58. Qiu, G. Y., Ben, A. J., Yano, T., & Momii, K. (1998). An improved methodology to measure evaporation from bare soil. *J. Hydrol.,* 210, 93–105.
59. Qiu, G. Y., Ben, A. J., Yano, T., & Momii, K. (1999). Estimation of soil evaporation using the differential temperature method. *Soil Sci. Soc. Am. J.,* 63, 1608–1614.
60. Ranatunga, K., Nation, E. R., & Barratt, D. G. (2008). Review of soil water models and their applications in Australia. *Environmental Modeling & Software, 23,* 1182–1206.
61. Richard, G. A., Pruitt, W. O., Hon, D. Raes, Smith, M., & Pereira, L. S. (1974). Estimating evaporation from bare soil and the crop coefficient for the initial period using common soils information. *J. Irrig. Drain. Eng.,* 131(1), 1.
62. Ritchie, J. T. (1972). Model for predicting evaporation from row crop with incomplete cover. *Water Resour. Res.,* 8, 1204–1213.
63. Ritchie, J. T., & Adams, J. E. (1974). Field measurement of evaporation from soil shrinkage cracks. *Soil Sci. Soc. Am. J.* 38, 131–134.
64. Ritchie, J. T., & Johnson, B. S. (1990). Soil and plant factors affecting evaporation. In: *Irrigation of Agricultural Crops,* B. A. Stewart, D. R. Nielsen, eds., Chap. 13, Ser. 30, American Society of Agronomy, Madison, Wis., p. 363–390.
65. Rushton, K. R., Eilers, V. H. M., & Carter, R. C. (2006). Improved soil moisture balance methodology for recharge estimation. *J. Hydrol.,* 318, 379–399.
66. Sharma, M. K., & Hassan, B. (1998). Estimation of pan evaporation using metrological parameters. *J. Hydrol.,* 18(3–4), 1–9.
67. Shriyangi, A., & Kunio, W. (2003). Measurement of evaporation on bare soil and estimating surface resistance. *J. Envir. Eng.,* 129(12), 1157–1168.
68. Singh, R. V., Chauhan, H. S., & Ali, A. B. M. (1981). Pan evaporation as related to metrological parameters. *J. Agril. Engg.,* 18(1), 48–53.
69. Snyder, R. L., Bali, K., Ventura, F., & Gomez-Macpherson, H. (2000). Estimating evaporation from bare or nearly bare soil. *J. Irrig. Drain. Eng.,* 126(6), 399–403.
70. Soares, J. V., Bernard, R., Taconet, O., Vidal-Madjar, D., Weill, A. (1988). Estimation of bare soil evaporation from airborne measurements. *J. Hydrol.,* 99, 281–296.

71. Stroosnijder, L. (1987). Soil evaporation: a test of a practical approach under semi-arid conditions. *Netherlands J. Agric. Sci.,* 34, 417–426.

72. Torres, E. A., Rubio, E., Calera, A., & Cuesta, A. (2006). Evaluation of FAO-56 model for bare soil evaporation in a semi-arid region using experimental data. *Geophysical Research Abstracts,* 8, 3632.

73. Ventura, F., Snyder, R. L., & Bali, K. M. (2006). Estimating evaporation from bare soil using soil moisture data. *J. Irrig. Drain. Eng.,* 132(2), 153–158.

74. Wallace, J. S., Jackson, N. A., & Ong, C. K. (1999). Modeling soil evaporation in an agroforestry system in Kenya. *Agric. For. Meteoro.,* 94, 189–202.

75. Wu, A., Black, T. A. K., Verseghy, D. L., Novak, M. D., & Bailey, W. (2000). Testing methods of estimating evaporation from bare and vegetated surfaces in CLASS. *J. Atmosphere-Ocean.,* 38 (1), 15–35.

76. Wythers, K. R., Lauenroth, W. K., & Paruelo, J. M. (1999). Bare-soil evaporation under semiarid field conditions. *Soil Sci. Soc. Am. J.* 63, 1341–1349.

77. Yunusa, I. A. M., Sedgley, R. H., & Tennant, D. (1994). Evaporation from bare soil in south-western Australia: a test of two models using Lysimetry. *Aust. J. Soil Res.,* 32, 437–446.

78. Zhi, G., Kai, Z., & Dong-sheng, S. (2006). Experimental study on soil moisture using dual-frequency microwave radiometer. *Chinese Geographical Science,* 16(1), 83–86.

CHAPTER 3

TREE INJECTION IRRIGATION: PRINCIPLES, PERSPECTIVES AND PROBLEMS[1]

MEGH R. GOYAL

CONTENTS

---

[1]This chapter is modified and translated from *Goyal, Megh R., E. A. González y O. Colberg Rivera. 1990. Riego al Xilema: Principios, Perspectivas y Problemas. Capitulo XX. En: Manejo de Riego por goteo editado por Megh R. Goyal, páginas 469–496. Río Piedras, PR: Servicio de Extensión Agrícola, UPRM.* For more details, one may contact by E-mail: goyalmegh@gmail.com. The English version was presented as ASAE paper.

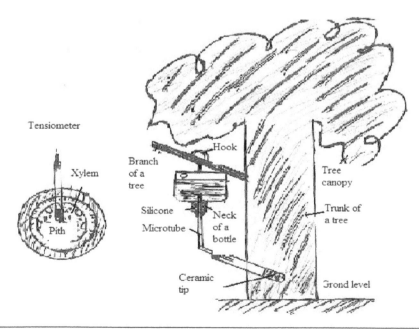

Tensiometer in a trunk tree

3.1   INTRODUCTION

With only a small portion of cultivated area under irrigation and with the scope to the additional area, which can be brought under irrigation, it is clear that the most critical input for agriculture today is water. It is accordingly a matter of highest importance that all available supplies of water should be used intelligently to the best possible advantage. Recent research around the world has shown that the yields per unit quantity of water can be increased if the fields are properly leveled, the water requirements of the crops as well as the characteristics of the soil are known, and the correct methods of irrigation are followed. Very significant gains can also be made if the cropping patterns are changed so as to minimize storage during the hot summer months when evaporation losses are highest, if seepage losses during conveyance are reduced, and if water is applied at the critical times when it is most useful for plant growth.

The main objective of irrigation is to provide plants with sufficient water to prevent stress that may reduce the yield. The frequency and quantity of water depends upon local climatic conditions, crop and stage of growth and soil-moisture- plant characteristics. Need for irrigation can be determined in several ways that do not require knowledge of evapotranspiration [ET] rates. One way is to observe crop indicators such as change of color or leaf angle, but this information may appear too late to avoid reduction in the crop yield or quality. Other similar methods of scheduling include determination of the plant water stress, soil moisture status or soil water potential. Methods of estimating crop water requirements using ET and combined with soil characteristics have the advantage of not only being useful in determining when to irrigate, but also enables us to know the quantity of water needed. ET estimates have not been made for the development countries though basic information on weather data is available. This has contributed to one of the existing problems that the vegetable crops are over irrigated and tree crops are under irrigated.

Water supply in the world is dwindling because of luxury use of under ground sources; competition for domestic, municipal and industrial demands; declining water quality (as happened in other drought areas); and losses through seepage, runoff, and evaporation. Water rather than land is one of the limiting factors in our goal for self-sufficiency in agriculture (e.g., Puerto Rico imported 70% of its food consumption from Dominican Republic and other countries). Intelligent use of water will avoid problem of sea water entering into aquifers. Introduction of new irrigation methods has encouraged marginal farmers to adopt these methods without taking into consideration economic benefits of conventional, overhead and drip irrigation systems. What is important is "net in the pocket" under limited available resources. Irrigation of crops in tropics requires appropriately tailored working principles for the effective use of all resources peculiar to the local conditions. Irrigation methods include border-, furrow-, subsurface-, sprinkler-, sprinkler, micro, drip/trickle and xylem irrigation.

3.1.1 DEFINITION OF TREE INJECTION IRRIGATION

Tree injection irrigation is a direct application of irrigation waters along with plant nutrients/growth regulators or inhibitors, pesticides or other chemicals into xylem of the tree trunk, by using series of injection sites depending upon the age of the tree. Tree injection irrigation is also called ultra-, micro-, high frequency-, tension-, tree injection-, trunk-, agro-stoichiometric-, sap-, internal circulatory-, or chemotherapy irrigation. There is no difference in the concept these names represent. The basic idea originated when various chemicals were injected into the internal circulatory system of tree. It is simple to inject water, fertilizers, micronutrients, growth promoters, growth inhibitors, pesticides, trace elements, gases, precursors of flavors/color and aroma, and in general any substance valuable for the improvement of fruit quality.

3.2 PRINCIPLES OF SOLUTE MOVEMENT IN TREES

3.2.1 STRUCTURE AND GROWTH OF THE TREE TRUNK

Dicot and monocot stems have number of structures and cell types in common, but have certain differences in the arrangements of their tissues (Figures 3.1–3.4). Both have an outer layer of *epidermis*, usually covered on the outside with waxy cuticle. The *parenchyma*, consists of large, thin walled, and relatively undifferentiated cells. Outside the vascular

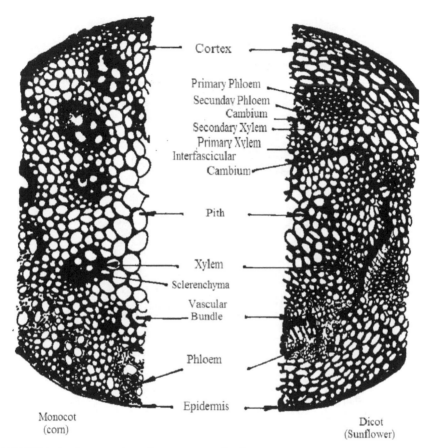

Cortex

Primary Phloem
Secunday Phloem
Cambium
Secondary Xylem
Primary Xylem
Interfascicular
Cambium

Pith

Xylem

Sclerenchyma

Vascular
Bundle

Phloem

Epidermis

Monocot
(corn)

Dicot
(Sunflower)

FIGURE 3.1 Cross section of a monocot and a dicot stem.

bundles is the *cortex*, usually composed of smaller, more differentiated parenchyma, and inside is the *pith*, composed of somewhat larger, thinner-walled parenchyma cells. The vascular bundles of monocots are scattered throughout the parenchyma, where as those of dicots are arranged in a ring. Each vascular bundle contains *xylem* cells toward the center and phloem toward the outside.

The *phloem* is composed mainly of large diameter, thin walled cells with characteristic sieve like end plates called *sieve elements* that are lined up end to end to make *sieve tubes*. These are associated with small parenchyma cells called companion cells. Vessels and tracheids die as they mature and lose their cell contents, but phloem cells, as well as the non-specialized parenchyma cells of the cortex and pith, stay alive and

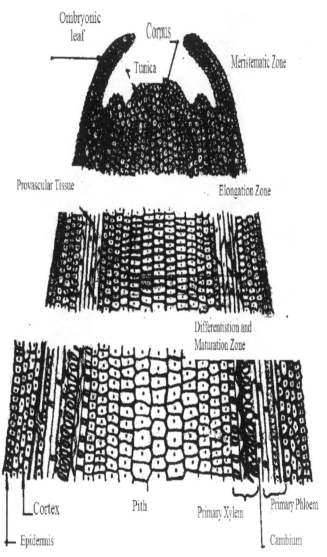

FIGURE 3.2   Cross section of a monocot and a dicot stem.

retain some of its structural integrity. Sieve elements may lose their nuclei and undergo extensive modifications in structure, but these stay alive and apparently are able to metabolize.

Xylem is primarily composed of dead, thick-walled conducting cells, either *vessels* (large cells with no cross walls, forming tube like pipes that

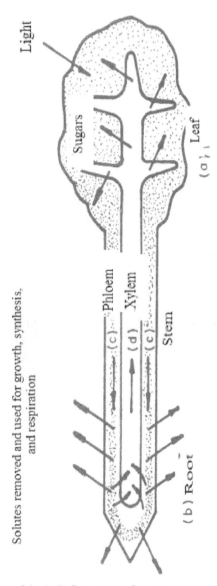

FIGURE 3.3 Diagram of the bulk flow system in a tree.

run longwise through the stem) or tracheids (much smaller in diameter, having end walls, and usually heavier secondary thickening). The xylem may contain fibers (similar to tracheids but with longer, narrower tips) that serve primarily for structural support. The strands or sheets of parenchyma

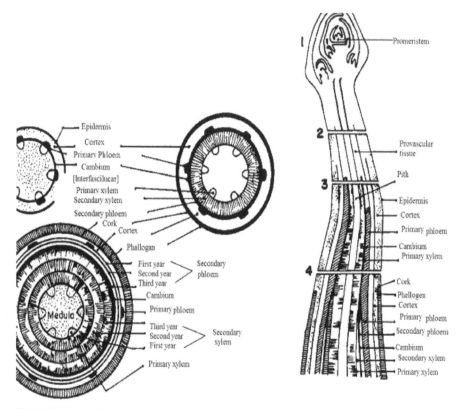

FIGURE 3.4 Cross sections of a woody stem with secondary growth. These cross sections can be related to the longitudinal section at the right.

cells penetrate the xylem. The main difference between monocot and dicot stems is in the organization of the bundles, and the existence of meristematic tissue in dicot bundles.

Monocots have bundles scattered throughout the parenchyma, each containing xylem toward the inside and phloem toward the outside. The first-formed xylem, called *proto-xylem* is nearest the center and later xylem, called *metaxylem*, is nearest the phloem. No cell division takes place once the bundles are formed. A large part of the maturation and differentiation of the tissue takes place before elongation of the monocot stem.

Dicot stems are more complex and almost invariably capable of secondary growth. Initially the bundles are arranged in a circle around a central core of pith. The xylem and phloem are separated by a layer of cells capable of division, called the *cambium*. Secondary growth occurs from this cambium by divisions that are tangential to the circumference of the stem, giving rise to new phloem cells to the outside and new xylem cells to the inside. Later, *interfascicular cambium* (inner fascicle, between bundles) develops by rejuvenation of parenchyma cells between the bundles.

Thus a complete circle of cambium is formed, which forms a circle of xylem to the inside and a circle of phloem to the outside. The *stele* is the whole central section of the stem, including the phloem and everything inside it. The outer cortex and the outer layers of phloem give rise periodically to cork cambium or *phellogen*, which produces the cork cells (phloem) that mainly constitute the bark. As the dicot stem enlarges in diameter, older bark sloughs off and new bark is formed from cork and the crushed layers of old phloem.

Perennial (woody) dicots may continue to expand for a prolonged period of time by secondary growth (Figure 3.4). Secondary xylem is deposited in annual rings containing large celled spring wood, which often contains the majority of vessels in woody angiosperms or hardwoods and smaller-celled summer wood. The perennial dicot stem seldom retains more than two years of phloem growth; the older phloem dies and sloughs off as the stem enlarges. The bases of branches are surrounded by new wood, forming knots in the wood.

3.2.2   XYLEM TRANSPORT

When the water is lost by transpiration, it must be replaced through the roots. Water loss from leaves implies the decrease in the amount of water in the plant, consequently its potential is lower (becomes more strongly negative) and water diffuses into roots down the potential gradient. The larger absorbing surface of the roots provides the necessary contact between the aerial part of the plant and the soil water.

Water can diffuse from cell to cell down a potential gradient and may enter the xylem with sufficient force to generate pressure as high as 2–3 bars or higher. However, this root pressure is never sufficient to raise the water to the top of a tall tree, and it may be very low most of the time, particularly at times when water loss is greatest. Furthermore, the flow from root pressure is not great enough to account for the volume of water that actually moves through a tree. It requires a pressure of 10 bars to raise water to a height of 300 feet, and more pressure is required to overcome resistance in the trunk and to maintain an adequate flow. Clearly this pressure cannot be supplied from below since root pressure of this magnitude has not been measured. Capillary and the matrix potential of a system with narrow passages may be sufficient to raise water a short distance in stems but not to the height or in the quantities required.

It may imply that water may move due to a potential gradient from soil to atmosphere via xylem. It means that a very low potential of water in the atmosphere, relative to the potential of water in the soil, supplies the force that moves water up the plant to the leaves. In other words, as water evaporates from the leaf surface, more water is "pulled" up by the tension created. It seems incredible that the fragile cells of the leaf could withstand tensions of 150 psi or greater without collapsing. In fact due to small size, cells can withstand much greater tensions. The water potential of air is very low. For about 50% relative humidity, it is close to [−1000] bars and it is much greater at low RH:

$$RH = [-10.7(\theta)] * \log [100/RH] \tag{1}$$

where, θ = temperature in degrees absolute, and RH = relative humidity in percent.

The water potential difference ["A"] between the leaf cells and the atmosphere is often very high, and water loss from leaf-cell surfaces induces tremendous tension inside the cells. It is relieved by the flow of water from internal cells, and ultimately from the xylem of leaf veins, so the tension is transmitted to the water in the xylem. *Cohesion of water*: The problem now is, how can water be pulled by a tube or pipe system, such as the xylem, for distances greater than the height of a water column supported by 1 atmosphere [30 feet]? If one tries to pull water up a pipe by applying suction at the top, the column will break and a vacuum will form for a column height greater than 30 feet. In fact water molecules have great affinity for each other, and narrow columns of water may withstand a tension up to 1000 atmospheres without breaking due to cohesive properties of the molecules. The theory that water could be pulled up very tall trees in long threadlike columns in the xylem was advanced in 1894.

Such columns do not normally cavitate because the cohesive property of water molecules is sufficient to prevent their pulling apart or pulling away from the walls of the vessels.

Evidence supporting the theory that water is pulled up by the forces of evaporation from the leaves is indirect. It has been clearly shown by injecting dyes, radioactive tracers, or small pulses of heat into a tree trunk and following the movement of marker that water does move in xylem.

The fact that the xylem water is under tension can be observed by cutting into the stem of a plant; water will snap back into the xylem, and if water is added to the cut surface it is pulled in. If a sensitive microphone is placed against the stem, the snapping of xylem water threads may be heard, particularly on dry hot days. Extensive cavitations may cause severe wilting, because a broken column can no longer transmit to the roots, the tension required to lift water. It is these broken threads that are presumed to be reunited by root pressure at night when the tension caused by evaporation of water from the leaves is low.

One problem with this theory of water movement in stems is that breaks in water column do occur as a result of excess drought, the formation of gas bubbles from dissolved gas; and from mechanical leaks. Such breaks ought to theoretically inactivate the xylem strand in which they occur and reduce the capacity of the system to move water. Many such breaks do occur but they do not appear to affect water movement seriously. Presumably when the tension is relaxed at night, the columns again rejoin. If cuts are made in the tree trunk so that no continuous vertical columns of xylem remain, water is still able to ascend in a zigzag path, at a reduced rate. It seems possible that a lateral transport, that must occur, takes place as a result of diffusion in the xylem parenchyma (living tissue of xylem).

3.2.2.1 Vessel Size

It has been shown that in small vessels, the flow rate of water varies with the vessel radius (Poiseuille's law):

$$Q = [\pi R^4 \, \Delta P]/[8 \, \mu L] \tag{2}$$

where, Q = volume flow rate in cubic liters per second, R = radius in m, ΔP = change in pressure, N/ m², L = length in m, and μ = absolute viscosity in N-s/m².

Plants with long narrow stems, such as vines, tend to have large diameter vessels that permit high flow rates. Since flow rate varies inversely as length, this arrangement permits the efficient transfer of water through long distances in a stem of small cross section. However, such vessels are more subject to cavitations or breakage of the water column and the high root pressure of vines may be associated with the need to refill vessels that were emptied by cavitations.

On the other hand, trees, with a much larger cross section in relation to their length, tend to have smaller xylem conducting elements. This means that water can be drawn to greater heights by greater forces, and the reduction in carrying capacity caused by smaller diameter vessels is offset by the increased diameter of the trunk and the correspondingly greater number of conducting elements.

Extremes of temperature change are likely to cause bubble formation in water under tension. Plants in colder temperature zones tend to have smaller vessel size than plants living in tropical zones. Coniferous trees have no vessels at all but only tracheids.

3.2.3   ALTERNATE THEORIES

The water may ascend trees largely as vapor. However, most of the water in the xylem is known to be liquid, not vapor. Various suggestions of active pumping systems have been made, but no mechanical or biochemical devices that could accomplish such pumping have been found. Active water transport in living cells of trees seems unlikely. First the resistance of living cells to flow would be very great and would add greatly to the forces required to move quantities of water. If roots are cut off a wilted plant and the stem is placed in water, the plant recovers for more quickly than when the roots of the intact plant are placed in water. It shows the much greater resistance to water transport offered by living tissue and suggest that water moves through the stem in nonliving tissue. The application of poisons to the stem of tree also has little or no effect on water movement, reinforcing the concept that nonliving cells are involved in water transport and that no energy input occurs in the stem itself. Thus the theory that water ascends in the xylem of high plants largely under the influence of forces of water evaporation from the leaf seems correct.

3.2.4 SOLUTE MOVEMENT

Since the xylem represents an open-ended water transport system with essentially one way movement, it seems likely that solutes move passively in the xylem by solvent-drag. They need not to move at the same rate as water, since their movements may be influenced by adsorption to the walls of the vessels or by diffusion down a potential gradient within the flowing system. Provided that some active transport system is available to transfer solutes to the xylem apoplast, in other words to load the translocation stream at the bottom, no motive power other than flow of water is required to move them to the top of the open xylem system. At this point solutes could be removed by active transport or by diffusion, depending on the concentration of solutes in the leaf cells and their requirements. It may be noted that:

1. Salts and inorganic substances move upward in the xylem and downward in the phloem.
2. Organic substances move up and down in the phloem.
3. Organic nitrogen may move up in the xylem (e.g., in trees) or phloem (herbaceous plants).
4. Organic compounds like sugar may be present in the xylem sap in large concentrations during the spring when the sap rises in trees before the leaves emerge.
5. Lateral translocation of solutes from one tissue to another occurs, presumably by – normal mechanisms of transfer (diffusion, active transport, and so on).

3.3 TREE INJECTION IRRIGATION: PRINCIPLES AND PRACTICES

3.3.1 ADVANTAGES

3.3.1.1 Efficient Use of Water

1. No evaporation from the soil surface into the atmosphere.
2. No infiltration into the subsoil where roots are incapable of absorbing moisture.

3. No runoff.
4. No wetting of foliage.
5. Inhibits non beneficial consumptive use of water by weeds because the terrain is free – of weeds.
6. One can irrigate the entire field up to edges.
7. Accurate quantity of irrigation water can be applied according to transpiration rate – of the plant.
8. Overall water application efficiency can go up to 99%.
9. Savings up to 95% of water use can be achieved.

3.3.1.2 Plant Response

1. Crop growth characteristics can be manipulated.
2. Better fruit quality and uniformity of crop is expected.

3.3.1.3 Root Environment

1. Shallow root system.
2. Effective soil aeration.
3. Provision of required amount of nutrients.

3.3.1.4 Pest and Diseases

1. Pesticides can be injected into the plant system.
2. Frequency of sprays can be reduced.
3. Reduction in incidence of insects and diseases.
4. Reduced application rates of pesticides.

3.3.1.5 Weed Growth

1. It is minimum.
2. No weeds in dry surface between trees.

3.3.1.6 Agronomical Benefits

1. Irrigation activities do not interfere with cultivation, spraying, picking and handling.

2. Less inter-cultivation, soil crusting, and compaction problems.
3. No surface runoff.
4. No soil/water erosion due to irrigation.
5. Fertigation system is possible thus savings in energy and quantity.
6. Other necessary chemicals can be applied along with irrigation water.

3.3.1.7 Engineering and Economic Benefits

1. Significant savings in energy.
2. Cost is low compared to surface sprinkler and drip irrigation systems.
3. Pipe sizes are significantly smaller compared to pipe sizes in other irrigation systems.
4. Conveyance efficiency and water use efficiency can be increased up to 99%.
5. System can be installed in uneven terrains.
6. It requires constant discharges at low pressures.
7. Water and chemical use can be programed with the crop response.

3.3.2 DISADVANTAGES

1. May not be applicable in vegetable crops as it is more convenient to inject into tree – trunks.
2. May not be used in monocot species as the xylem is not as differentiated.
3. Introduction of new substances can cause toxic effects just as a man can overdose on – drugs.
4. May cause fungus growth at the injection site.
5. Holes in tree trunk must be made to install injection tips thus causing physical injury – to the plant.

3.3.3 OPERATIONAL PROBLEMS

1. No information is available on number of injection sites, water application rates, dosages of various chemicals depending upon the age of the tree.

2. At what height and depth, should the injection points be located?
3. Injection tips can be easily clogged by gum, wax and resins of tree.
4. Effective cleaning agent needs to be found to avoid clogging of tips.
5. Algae formation in the injector lines.
6. Laterals may contain air [from the tree] and thus obstructing the flow.
7. Leakage of water at the contact point between the injector tip and tree surface.
8. Excess pressure might loosen the sealing agent [silicon] and may throw out the – injection tip.
9. Expert advice is needed to locate xylem.
10. Chemigation might disturb the osmotic and electrical internal equilibrium in the plant.
11. Screening of pesticides and chemicals suitable for xylem irrigation.
12. Salts in excess of 300–500 ppm may require desalinization of water.
13. A clean, pure or soil water is necessary.

3.3.4 PRINCIPLE OF OPERATION

It is based on utilizing natural negative sap pressures within a plant to suction liquids and gases directly into the inner circulatory system, analogous to a human blood transfusion. The technique is accomplished by placing an injection tip (e.g., a ceramic implant) directly in the xylem layer, the negative pressure area. Liquid or gas is then made available to the implant through a plastic tubing at very little or no pressure. Fluids can then traverse in the plant in any direction. The roots of the plant continue to be nourished by the natural way, with sap, water and nutrients. The roots still seek moisture and grow down using a stimulus called geotropism.

Plants give off water through a process called transpiration. The amount of water a plant "throws off" and the amount it needs are two different situations. A well-known "Hill Reaction" is defined below:

$$6\ CO_2 + 6\ H_2O + H_2 = C_2H_{12}O_6 + 6 - O_2 \tag{3}$$

Opposite of transpiration is called respiration. By careful measurement of the quantities of sugar synthesized in the leaves by unitary surface and time (10–15 mg of hexose/dm^{2}h), it is readily calculable what stoichiometric quantity of water is required under the same conditions (e.g., 50–80 ml for a period of 8 hrs considering a canopy surface from 8–10 m^2). This quantity is very approximate to the quantity of water consumed by xylem irrigation during the same period under the same conditions. Primary water uptake occurs only during photosynthesis or day light [12].

3.3.5   SYSTEM MODIFICATION

It is accomplished by simply placing the ceramic piece in the root zone of house or commercial indoor plants, nursery stock or almost any plant too small to receive an implant in the trunk. The same efficient use of water and nutrients are applicable but some of the metabolic engineering techniques (modulation of the plant metabolism with the aim of obtaining better fruits by injection of substances such as promoters of color, bouquet, flavor, aroma, metabolites, enzymes or coenzymes) may not be effective. Seeds for greenhouses can also be germinated and grown from an implant in the soil. The seed can actually be glued to the implant, then planted and grown through maturity. Tree crops can be raised with other irrigation systems and then xylem tips can be installed after first year of growth. Water usage of 40 ounces/day on older trees, 5 ounces/day for grapes has been reported. This calculates to be approximately 0.05 gpm/acre of irrigation during 12 hours photosynthesis or 36 gallons/day/acre.

3.4   DESCRIPTION OF TREE INJECTION IRRIGATION SYSTEM

The system consists of a water resource, pump, chemigation system, filter, main line, sub main, laterals, and injection tips. The installation of injection tip should be done in the following manner:

1.   Select the size of a ceramic tip.
2.   Select the best location on the tree trunk.

3. Bore a hole through the cambium layer approximately 1/4 diameter larger than the – injector.
4. Use a sharp instrument to remove the plug of bark. It is important to bore past the – phloem to cause leakage out of the plant.
5. Continue the bore into xylem (sapwood) portion to the same dimension as length of – ceramic portion of the injector. The hole should allow a snug fit.
6. Use an inert sealing agent (silicone) for sealing the injector to tree.
7. Hook water to be injected to the tip at a pressure of 1 to 8 ft. of head (necessary – pressure can be allowed by gravity, low-pressure pump).
8. Very minute quantities of chemical can be injected into the water stream using a plastic syringe (doctor's needle).

3.4.1 PRECAUTIONS FOR TREE INJECTION IRRIGATION SYSTEM

1. Sterilize the drill bit.
2. Hole should allow a perfect fit.
3. Use a good sealing agent.
4. Water should be free from pathogens.
5. Use a pesticide to avoid fungal growth.
6. Injector site should be allowed to dry before starting irrigation.
7. No leakage can be allowed between the tree and tip, as it will break the suction.
8. High precaution is essential in determining dosage of the chemicals to avoid toxic – hazards in the plant.
9. Any injection holes, which cannot be used, should be left open. They heal with time.

3.5 CURRENT STATUS

Table 3.1 reveals current information available in different areas of tree injection irrigation. Most of the information is available on injection of pesticides into tree trunk using syringe, plastic bottle by gravity, etc.

TABLE 3.1 Current Status for Tree Injection Irrigation

Description	References
A. **Chemigation:**	5, 8, 14, 15, 16, 20, 22, 25, 26, 34, 35
1. Fertilizer/micronutrients	7, 38
2. Growth regulator/inhibitor	2, 3, 21, 29, 33, 42, 44, 46
3. Pesticide	4, 6, 13, 18, 19, 20, 24, 27, 36, 37, 39, 40, 41, 43, 46, 49
B. **Irrigation:** technology and principles:	12
C. **Injection** principles and technology:	5, 20, 32, 42, 48, 50
1. Equipment	2, 4, 17, 19, 25, 28, 31
2. Injectors	19, 39, 45, 50
3. Methods	5, 6, 7, 8, 13, 17, 18, 25, 26, 28, 33, 35
4. Pressure injection	17, 19, 22, 29, 31, 34, 38, 41, 42, 43, 45
D. **Measurement of xylem water potential**	1, 23

Note: Numbers refer to appended references.

The information on tree injection irrigation needs to be developed. This offers new challenges for researchers and students, who want to pursue MSc/PhD/Postgraduate diploma/Post-doc. The opportunities are infinite. The 1984 Spring issue of Drip/Trickle Irrigation describes briefly the technology involved. Table 3.2 reveals measured potentials in different parts of the tree. Monthly mean intensity of solar radiation is given in Table 3.3. The percentage of daily sunshine hours is indicated in Table 3.4. Differences between dicot and monocot trees are detailed in Appendices 1 and 2.

3.6 PRELIMINARY STUDIES

To establish a basis for irrigation scheduling in tree injection irrigation, the tensiometers were installed in the soil and tree trunk in drip irrigated, flood irrigated and micro sprinkler irrigated mango trees. Tensiometers were read for 13 days at 7:00 a.m., 12:00 Noon and 3:00 p.m. on each day. In all cases, it was observed that the tree tensiometer did not respond

TABLE 3.2   Actual Measured Water Potentials in Soil and Different Parts of Trees at Various Times

Parameter	Species						
	Juniper us scopulorum (juniper)				*Ulmus parvifolia* (elm)		
	September						
Date	17	18	20	22	12	16	17
Time, hr	1500	500	200	1500	1400	1400	400
Conditions	Clear	night	rain	clear	clear	Clear (irrigation)	night
Temperature, °C	18	10	12	11	27	16	7
ψ Soil, bars	-5.7	-6	-7.1	-0.2	-4.6	-0.1	-0.1
ψ trunk, bars	-8.6	-6.6	-7.9	-5	-7	-2.6	-2.6
ψ branches, bars	-12	-9.4	-8.7	-8	-7.6	-5.3	-4.1
ψ twigs, bars	-21.8	-11.8	-12.2	-15.6	-23	-16.8	-5.7
ψ leaves, bars	-40	-25	------	------	-24.5	-23.9	-10.7

Parameter	Species			
	Elaeagnus angustifolia (Russian olive)		*Acer glabrum* (maple)	
	September			
Date	16	17	17	18
Time,hr	1500	500	1500	500
Conditions	clear	night	clear	night
Temperature, °C	16	7	16	7
ψ Soil, bars	-3.3	-3.3	-5.7	-6
ψ trunk, bars	-7.4	-3.7	-8.9	-6.8
ψ branches, bars	-----	------	------	------
ψ twigs, bars	-17.7	-10.7	-25.2	-8.8
ψ leaves, bars	-31.9	-17.9	-43	-32.2

Source: H. H. Wieber, R. W. Brown, T. W. Daniel and E. Campell. Bioscience. 20:226 [1970].

after few days of operation. This implied that the tips might get clogged. The variations in tension readings for tree tensiometers followed the same pattern as soil tensiometers. Soil tension readings were more pronounced as shown in Figure 3.5.

TABLE 3.3   Monthly Mean Intensity of Solar Radiation [Ra], Which is Measured Over a Horizontal Surface, in Millimeter of Water Evaporated Per Day (Following Blaney – Criddle)

Northern Hemisphere

Month	90°	80°	70°	60°	50°	40°	30°	20°	10°
Jan	—	—	—	1.3	3.6	6	8.5	10.8	12.8
Feb	—	—	1.1	3.5	5.9	8.3	10.5	12.3	13.9
March	—	1.8	4.3	6.8	9.1	11	12.7	13.9	14.8
April	7.9	7.8	9.1	11.1	12.7	13.9	14.8	15.2	15.2
May	14.9	14.6	13.6	13.6	15.4	15.9	16	15.7	15
June	18.1	17.8	17	16.5	16.7	16.7	15.8	14.8	13.4
July	16.8	16.5	15.8	15.7	16.1	16.3	16.2	15.7	14.8
Aug	11.2	10.6	11.4	12.7	13.9	14.8	15.3	15.3	15
Sept	2.6	4	6.8	8.5	10.5	12.2	13.5	14.4	14.9
Oct	—	0.2	2.4	4.7	7.1	9.3	11.3	12.9	14.1
Nov	—	—	0.1	1.9	4.3	6.7	9.1	11.2	13.1
Dec	—	—	—	0.9	3	5.5	7.9	10.3	12.4

Southern Hemisphere

Month	0°	10°	20°	30°	40°	50°	60°	70°	80°	90°
Jan	14.5	15.8	16.8	17.3	17.3	17.1	16.6	16.5	17.3	17.6
Feb	15	15.7	16	15.8	15.2	14.1	12.7	11.2	10.5	10.7
March	15.2	15.1	14.6	13.6	12.2	10.5	8.4	6.1	3.6	1.9
April	14.7	13.8	12.5	10.8	8.8	6.6	4.3	1.9	—	—
May	13.9	12.4	10.7	8.7	6.4	4.1	1.9	0.1	—	—
June	11.6	9.6	7.4	5.1	2.8	0.8	—	—	—	—
July	13.5	11.9	10	7.8	5.6	3.3	1.2		—	—
Aug	14.2	13	11.5	9.6	7.5	5.2	2.9	0.8	—	—
Sept	14.9	14.4	13.5	12.1	10.5	8.5	6.2	3.8	1.3	—
Oct	15	15.3	15.3	14.8	13.8	12.5	10.7	8.8	7.1	7
Nov	14.6	15.7	16.4	16.7	16.5	16	15.2	14.5	15	15.3
Dec	14.3	15.8	16.9	17.6	17.8	17.8	17.5	18.1	18.9	19.3

Source: Shaw, N. 1986. *Manual of Meteorology*. Volume II. Second edition. Pages 4–5. Cambridge University Press.

Note: The values in the table have been multiplied by 0.86 and divided by 59 to calculate radiation in millimeters of water per day.

TABLE 3.4 Percentage of Daily Sunshine Hours for North and South Hemisphere

Latitude	Jan	Feb	Mar	Apr	May	June	July	Aug	Sept	Oct	Nov	Dec
Northern Hemisphere												
0	8.5	7.76	8.49	8.21	8.5	8.22	8.5	8.49	8.21	8.5	8.22	8.5
5	8.32	7.57	8.47	8.29	8.65	8.41	8.7	8.6	8.23	8.42	8.07	8.3
10	8.13	7.47	8.45	8.37	8.81	8.6	8.9	8.71	8.25	8.34	7.91	8.1
15	7.94	7.36	8.43	8.44	8.98	8.8	9.1	8.83	8.28	8.2	7.75	7.88
20	7.74	7.25	8.41	8.52	9.15	9	9.3	8.96	8.3	8.18	7.58	7.66
25	7.53	7.14	8.39	8.61	9.33	9.23	9.5	9.09	8.32	8.09	7.4	7.42
30	7.3	7.03	8.38	8.72	9.53	9.49	9.7	9.22	8.33	7.99	7.19	7.15
32	7.2	6.97	8.37	8.76	9.62	9.59	9.8	9.27	8.34	7.95	7.11	7.05
34	7.1	6.91	8.36	8.8	9.72	9.7	9.9	9.33	8.36	7.9	7.02	6.92
36	6.99	6.85	8.35	8.85	9.82	9.82	10	9.4	8.37	7.85	6.92	6.79
38	6.87	6.79	8.34	8.9	9.92	9.95	10	9.47	8.38	7.8	6.82	6.66
40	6.76	6.72	8.33	8.95	10	10.1	10	9.54	8.39	7.75	6.72	7.52
42	6.63	6.65	8.31	9	10.1	10.2	10	9.62	8.4	7.69	6.62	6.37
44	6.49	6.58	8.3	9.06	10.3	10.4	10	9.7	8.41	7.63	6.49	6.21
46	6.34	6.5	8.29	9.12	10.4	10.5	11	9.79	8.42	7.57	6.36	6.04
48	6.17	6.41	8.27	9.18	10.5	10.7	11	9.89	8.44	7.51	6.23	5.86
50	5.98	6.3	8.24	9.24	10.7	10.9	11	10	8.46	7.45	6.1	5.65
52	5.77	6.19	8.21	9.29	10.9	11.1	11	10.12	8.49	7.39	5.93	5.43
54	5.55	6.08	8.18	9.36	11	11.4	11	10.26	8.51	7.3	5.74	5.18

TABLE 3.4 Continued

Latitude	Jan	Feb	Mar	Apr	May	June	July	Aug	Sept	Oct	Nov	Dec
56	5.3	5.95	8.15	9.45	11.2	11.7	12	10.4	8.53	7.21	5.54	4.89
58	5.01	5.81	8.12	9.55	11.5	12	12	10.55	8.55	7.1	4.31	4.56
60	4.67	5.65	8.08	9.65	11.7	12.3	12	10.7	8.57	6.98	5.04	4.22

Southern Hemisphere

Latitude	Jan	Feb	Mar	Apr	May	June	July	Aug	Sept	Oct	Nov	Dec
0	8.5	7.66	8.49	8.21	8.5	8.22	8.5	8.49	8.21	8.5	8.22	8.5
5	8.68	7.76	8.51	8.15	8.34	8.05	8.3	8.38	8.19	8.56	8.37	8.68
10	8.86	7.87	8.53	8.09	8.18	7.86	8.1	8.27	8.17	8.62	8.53	8.88
15	9.05	7.98	8.55	8.02	8.02	7.65	8	8.15	8.15	8.68	8.7	9.1
20	9.24	8.09	8.57	7.94	7.85	7.43	7.8	8.03	8.13	8.76	8.87	9.33
25	9.46	8.21	8.6	7.84	7.66	7.2	7.5	7.9	8.11	8.86	9.04	9.58
30	9.7	8.33	8.62	7.73	7.45	6.96	7.3	7.76	8.07	8.97	9.24	9.85
32	9.81	8.39	8.63	7.69	7.36	6.85	7.2	7.7	8.06	9.01	9.33	9.96
34	9.92	8.45	8.64	7.64	7.27	6.74	7.1	7.63	8.05	9.06	9.42	10.08
36	10.03	8.51	8.65	7.59	7.18	6.62	7	7.56	8.04	9.11	9.51	10.21
38	10.15	8.57	8.66	7.54	7.08	6.5	6.9	7.49	8.03	9.16	9.61	10.34
40	10.27	8.63	8.67	7.49	6.97	6.37	6.8	7.41	8.02	9.21	9.71	10.49
42	10.4	8.7	8.68	7.44	6.85	6.23	6.6	7.33	8.01	9.26	9.82	10.64
44	10.54	8.78	8.69	7.38	6.73	6.08	6.5	7.25	7.99	9.31	9.94	10.8
46	10.69	8.86	8.7	7.32	6.61	5.92	6.4	7.16	7.96	9.37	10.07	10.97

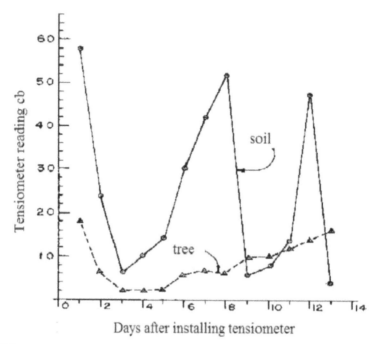

FIGURE 3.5   Tension readings based upon the tensiometer installed at 12 cm depth in the soil and tensiometer installed in the xylem of drip irrigated mango tree. Reading was recorded at 3:00 pm daily.

3.7   SUMMARY

The main objective of irrigation is to provide trees with the correct amount of water to prevent stress that can reduce yield. The frequency and quantity of water depends on the climatic conditions, the type of crop, and the stage of growth and soil-moisture-plant status. With only a small portion of cultivated area and the hope to have additional area under irrigation, it is clear that water is critical for agriculture and it is of vital importance.

Recent research has shown that the yields per unit quantity of water can be increased if we properly level the fields, if we know the crop requirements, the soil characteristics and if we use the correct methods of irrigation. Water supply is dwindling because of the excessive use of underground sources; domestic and industrial demands; declining water quality; and losses through seepage, runoff and evaporation. Water is one of the limiting factors in our goal for self-sufficiency in agriculture.

The introduction of new irrigation methods has encouraged farmers to adopt these methods without considering the economical benefits of conventional, overhead and drip irrigation systems. What is really important is "net in the pocket" under limited available resources. Some irrigation methods are border-, furrow-, subsurface, sprinkler-, micro sprinkler- drip/ trickle and xylem irrigation.

Tree injection (xylem) irrigation is a direct application of water along with nutrients, growth regulators or inhibitors, pesticides or other chemicals into the xylem of the tree trunk, by using a series of injection sites depending upon the age of the tree. The advantages of tree injection irrigation are: if water is used efficiently, one can save up to 95%. It prevents weeds from appearing because of the dry surfaces between the trees. Crop growth and better fruit quality can be expected. Pesticides can be injected into the plants reducing the frequency of application and the reduction of insects and diseases. Irrigation operations do not interfere with cultivation, spraying, picking and handling. Farmers can use fertilizers efficiently. Other necessary chemicals can be applied through with irrigation water. There are significant savings in energy and money.

Some disadvantages are: It may not be applicable in vegetable crops. It may not be used in monocot species. Introduction of new substances can cause toxic effects. It may cause fungus to grow at the injection sites. Holes in tree trunk must be made to install injection tips thus causing physical injury to the plant.

KEYWORDS

- chemigation
- dicot
- drip irrigation
- fertilizers
- flood irrigation
- monocot
- pesticide

- phloem
- relative humidity
- tensiometer
- tree
- tree crop
- tree injection
- tree injection irrigation
- vegetable crop
- xylem
- xylem irrigation

REFERENCES

I. Citations Used in This Chapter

1. Bidwel, R. G. S. (1979). *Plant Physiology.* Second edition. MacMillan Publishing Co., Inc. New York. Pp. 304–22.
2. Brown, G. K. (1978). Prototype equipment for commercial pressure injection of aqueous growth regulators into tress. *Arboriculture, J.,* 4, 7–13.
3. Brown, G. K., Kwokek, W. F., Wuertz, D. E., G. A Jumper, Wilson, C. L., & Carr, S. R. (1977). Regrowth reduction in American elm and sycamore by growth regulator injection. *J. Am Soc. Hort. Sci.,* 102, 748–51.
4. Buitemdag, C. H., & Bronkhorst, G. J. (1980). Further notes on injection of insecticides into citrus tree trunks – for the control of red scale and citrus psylla. *The citrus and Subtropical Fruit Journal,* November.
5. Carter, R. H. (1939). Chemicals and methods used in treatments of trees by injections. *U.S. Bur. Entomology, Plant Quar.,* E-467, pp. 25.
6. Cohen, M. (1974). Diagnosis of young tree decline, blight and sand hill decline of citrus by measurement of water up take using gravity injection. *Plant Dis. Bptr.,* 58, 801–805.
7. Collison, R. C., Harlan, J. D., & Sweeney, M. P. (1932). Direct tree injection in the study of tree nutrition problems. *N.Y. State Agr. Exp. Sta. Tech. Bull.* 192. Pp. 32.
8. Craighead, F. C., & St. George, R. A. (1938). Experimental work with the introduction of chemicals into the sap stream of trees for the control of insects. *Forestry, J.,* 36, 24–34.
9. Darvas, J. M., Toerien, J. C., & Milne, D. L. (1983). Trees for the effective control of phytophthora root rot. *The Citrus and Subtropical Journal,* March. Pp. 7–10.

10. Edwards, W. R. N., & Jarvis, P. G. (1981). A new method of measuring water potential in tree stems by water injection. *Plant Cell an Environment*, 4(6), 463–465.
11. Filer, Jr., T. H. (1973). Pressure apparatus for injecting chemicals into trees. *Plant Disease Report*, 57(4), 338–341.
12. Futuristic plant nutrition system will put profit back in farming. *Drip/Trickle Irrigation*, 6(1), 2–10.
13. Geyer, W. A. (1979). Tree control with injection of Garlon and Tordon brand herbicides. *Down to Earth*, 35, 9–13.
14. Goyal, M. R. (1980). Bibliography-Drip/Trickle Irrigation. NAPA Bull. No. 3, *National Agricultural Plastics Association*, Salisbury, MD.
15. Goyal, Megh, R., & Rivera, L. E. (1985). Bibliography – Drip/Trickle Irrigation (Supplement). Publication No. 84–13 by College of Engineers and Surveyors of Puerto Rico, GPO Box 3845, San Juan, P. R., 00936–3845, USA.
16. Goyal, M. R. (1985). CBAG – TAD – P. R. project proposal, "Irrigation Requirement Estimations in Puerto Rico." Agricultural Experiment Station, UPR-RUM, Rio Piedras.
17. Gregory, G. F., & Jones, T. W. (1975). *An Improved Apparatus for Pressure – Injecting Fluid Into Trees*. USDA Forest Service Research Note, NE- Z14.
18. Harries, F. H. (1965). Control of insects and mites on fruit trees by trunk injections. *Journal of Economic Entomology*, 58(4), 631–634.
19. Hedden, S. L., Lee, R. F., Timmer, L. W., & Albrigo, L. W. (1981). Improved injector design for pressure injecting citrus trees with tetracycline. *Phytopathology*, 71, 880.
20. Helburg, L. W., Schomaker, M. E., & Morrow, R. A. (1973). A new trunk injection technique for systemic chemicals. *Plant Disease Report*, 57(6), 513–14.
21. Hield, H., Boswell, S. B., & Hemstreet, S. (1977). Eucalyptus tree growth control by inhibitors applied as sprays, injection, cut painting or trunk banding. *J. Am. Soc. Hort. Sci.*, 102, 665–669.
22. Himelick, E. B. (1972). High pressure injection of chemicals into trees. *Arborist's News*, 37, 97–103.
23. Huzulak, J., & Matejka, F. (1980). Study of xylem pressure potential daily dynamics by means of autocorrelation analysis. *Biologia Plantarum*, 22(5), 336–340.
24. Jeppson, L. R., Jesser, M. J., & Complin, J. O. (1952). Tree trunk application as a possible method of using systemic insecticides on citrus. *J. Econ. Entomology*, 4, 669–671.
25. Jones, T. W., & Gregory, G. F. (1971). An apparatus for pressure injection of solutions into trees. *U. S. Forest Service Res. Pap. NE-233*. Pp. 7.
26. Lantz, A. E. (1938). An efficient method for introduction of chemicals into living trees. *US Bur. Ent. Plant, Q. E-434*. Pp. 4.
27. Lee, R. F., Timmer, L. W., & Albrigo, L. G. (1982). Effect of Oxytetracycline and Benzimidazole treatments on blight-affected citrus trees. *J. Am. Soc. Hort. Sci.*, 107(6), 1133–1138.
28. May, C. (1941). Methods of tree injection. *The Journal of General Physiology*, 4, 7,10–12,14,16.
29. Miller, S. S. (1982). Growth and branching of apple seedlings as influenced by pressure – injected plant growth regulators. *Hort. Sci.*, 17(5), 775–776.
30. Pryor, A. (1984). How to fight phytophthora in avocados. *Agrichemical Age*, March. Pp. 53–54.

31. Reil, W. O., & Bentel, J. A. (1976). A pressure machine for injecting trees. Calif. Agric., 30(12), 4–5.
32. Roach, W. A. (1939). Plant injection as a physiological method. Ann. Bot., 3, 156–225
33. Roberts, B. R., Wuertz, D. E., Brown, G. K., & Kwolek, W. F. (1979). Controlling sprout growth in shade trees by trunk injection. J. Am. Soc. Hort. Sci., 104, 883–887.
34. Sachs, R. M., Nyland, G., Hackett, W. P., Coffelt, J., DeBie, J., & Giannini, G. (1977). Pressurized injection of aqueous solutions into tree trunks. Scientia Hort., 6, 297–310.
35. Schreiber, L. R. (1969). A method for the injection of chemicals into trees. Plant Dis. Reporter, 5387, 64–65.
36. Schwarz, R. E., Moll, J. N., & Van Vuuren, S. P. (1976). Control of citrus greening and its psylla vector by trunk injections of tetracyclines and insecticides. Proc. 7th Conf. Int. Organ, Citrus Visol, Univ. of CA, Riverside- CA. pp. 26–29.
37. Schwarz, R. E., & Van Vuuren, S. P. (1971). Decrease in fruit greening of sweet orange by trunk injection of tetracyclines. Plant Disease Reporter, 55, 747–750.
38. Southwick, R. W. (1945). Pressure injection of iron sulfate into citrus trees. Proc. Am. Soc. Hortic. Sci., 46, 27–31.
39. Sterrett, J. P. (1968). Response of oak and red maple to herbicides applied with an injector. Weed Sci., 16, 159–160.
40. Sterrett, J. P. (1969). Injection of hardwoods with dicamba, picloram and 2,4-D. Journal Forestry, 67, 820–821.
41. Sterrett, J. P. (1979). A pressure injector for herbicidal control of potential danger trees. Proc. Northeast Weed Sci. Soc., 33, 207–212.
42. Sterrett, J. P. (1979). Injection methodology for evaluating plant growth retardants. Weed Sci., 27, 688–690.
43. Sterrett, J. P. (1982). Selective control of trees by pressure injection of herbicides. Hort. Science, 17(3), 360–361.
44. Sterrett, J. P. (1985). Paclobutzrazol: A promising growth inhibitor for injection into woody plants. J. Am. Soc. Hort. Science, 110(1), 4–8.
45. Sterrett, J. P., & Creager, R. A. (1977). A miniature pressure injector for deciduous woody seedlings and branches. Hort. Science, 12(2), 156–158.
46. Sterrett, J. P., Hodgson, R. H., & Snyder, R. H. (1983). Growth retardant response of bean and woody plants to injections of MBR 18337. Weed Sci., 31, 431–435.
47. Timmer, L. W., Lee, R. R., Albrigo, L. G. (1982). Distribution and persistence of trunk injected oxytetracycline in blight affected and healthy citrus. J. Am. Hort. Sci., 107, 428–432.
48. Tree injection – News. Farm Chemicals, 138(12), 64–66.
49. Van Vuuren, S. P., Moll, J. N., & de Graca, J. V. (1977). Preliminary report on extended treatment of citrus greening with tetracycline hydrochloride by tree injection. Plant Disease Reporter, 61, 358–359.
50. Young, R. H., Garnsey, S. M., & Horanic, G. (1979). A device for infusing liquids into the outer xylem vessels of citrus trees. Plant Disease Reporter, 63(9), 713–715.
51. Wieber, W, R., Brown, Daniel, T. W., & Campbell, E. (1970). Bioscience, 20, 226.

II. Books/Bulletins/Journal and Proceedings/Reports

52. Aajoud, A, Raveton, M., Aouadi, H., Tissut, M., & Ravanel, P. (2006). Uptake and xylem transport of fipronil in sunflower. *J. Agric Food Chem.*, 54(14), 5055–5060.
53. Benova, A, Digonnet, C., Goubet, F., Ranocha, P., Jauneau, A., & Pesquet, E. (2006). Galactoglucomannans increase cell population density and alter the protoxylem/ metaxylem tracheary element ratio in xylogenic cultures of Zinnia Elegans. *J. Plant Physiol.*, 56(11), 454–466.
54. Camp, R, C. (1999). Subsurface drip irrigation: A Review. *Transactions of the ASAE*, 41(5), 1353–1367.
55. Chatelet, D, Matthews, M., & Rost, T. (2006). Improvement of seed yields under boron-limiting conditions through overexpression of BOR1, a boron transporter for xylem loading, in Arabidopsis thaliana. *J. Plant Physiol.*, 56(6), 184–191.
56. Else, M, Taylor, J., & Atkinson, J. (2006). Anti-transparent activity in xylem sap from flooded tomato (Lycopersicon esculentum Mill.) plants is not due to pH- mediated redistributions of root- or shoot-sourced ABA. *J. Exp Bot.*, 71(15), 845–861. August.
57. Fekersillassie, D., & Eisenhauer, D. E. (2000). Feedback-controlled surge irrigation: I. Model development. *Transactions of the ASAE*, 43(6), 1621–1630.
58. Fekersillassie, D., & Eisenhauer, D. E. (2000). Feedback-controlled surge irrigation: II. Operating criteria. *Transactions of the ASAE*, 43(6), 1631–1641.
59. Green, Steve, Brent Clothier, & Jardine Bryan, (2003). **Theory and practical application of heat pulse to measure sap flow.** SCI Journals, 95: 1371–1379.
60. Hopkins, William, G., & Norman, Huner, P. A. (2002). *Introduction to Plant Physiology*. John Wiley and Sons. Pp. 1–576.
61. Ieperen, W., & Gelder, O. (2006). Ion-mediated flow changes suppressed by minimal calcium presence in xylem sap in *Chrysanthemum* and *Prunus laurocerasus. J. Exp. Bot.*, July, 68(11), 743–750.
62. Irmak, S., Haman, D. Z., Irmak, A., Jones, J. W., Campbell, J. W., & Yeager, K. L. (2003). New irrigation-plant production system for water conservation in ornamental nurseries: Qualification and evaluation of irrigation, runoff, plant biomass, and irrigation efficiencies. *Applied Engineering in Agriculture*, 19(6), 651–655.
63. Junghans, U, A. Polle A, Duchting, P., Weiler, E., Kuhlman, B., Gruber, F., & Teichman, T. (2006). Adaptation to high salinity in poplar involves changes in xylem anatomy and auxinphysiology. *J. Plant Cell Environ.*, 49(8), 519–531.
64. Kramer, E. (2006). How far can a molecule of weak acid travel in the apoplast or xylem? *J. Plant Physiol.*, 56(11), 433–436. August.
65. Laschimke, R, Burger, M., & Vallen, H. (2006). Acoustic emission analysis and experiments with tension. *J. Plant Physiol.*, 56(10), 325–342. July.
66. Li, J., Zhang, J., & Rao, M. (2005). Modeling of water flows and nitrate transport under surface drip fertigation. *Transactions of the ASAE,* 48(2), 627–637.
67. Maw, B. W., Young, J. R., & Marti, L. R. (1999). Evaluating a contaminant using a center pivot irrigation system. *Transactions of the ASAE*, 42(3), 799–802.
68. Meshkat, M., Warner, R. C., & Workman, S. R. (2000). Evaporation reduction potential in an undisturbed soil irrigated with surface drip and sand tube irrigation. *Transactions of the ASAE*, 43(1), 79–86.

69. Miwa, K, Takano, J., & Fujiwara, T. (2006). Xylem structure and connectivity in Grapevine (Vitis vinifera) shoots provides a passive mechanism for the spread of bacteria in Grape. *J. Plant Physol.*, 56(12), 583–594. September.
70. Mmolawa, K., & Or, D. (2000). Water and solute dynamics under a drip irrigated experiments and analytical model. *Transactions of the ASAE*, 43(6), 1597–1608.
71. Pastor, P. (2006). Nuevos criterios para la ciencia del riego plantean los avances en fisiología vegetal. *J. Plant Physiol.*, 56(13), 625–658. October.
72. Pate, John, S., Craig, Atkins, A., Kathy Hamel, David, McNeil, L., & David, Layzell, B. (2002). Transport of organic solutes in phloem and xylem of a nodulated legume. *Plant Physiology*, 63, 1082–1088.
73. Schneider, A. D., & Howell, T. A. (1999). Methods, amounts, and timing of sprinkler Irrigation for winter wheat. *Transactions of the ASAE*, 40(1), 137–142.
74. Volker Stiller John, S., & Uwe, Hacke, G. (2003). **Xylem hydraulics and the soil–plant–atmosphere continuum: opportunities and unresolved issues.** *SCI Journals*, 95, 1362–1370.
75. Windt, C, Vergeldt, F., & Jager, P. (2006). MRI of long-distance water transport: a comparison of the phloem and xylem flow characteristics and dynamics in poplar, castor bean, tomato and tobacco. *J. Plant Cell Environ.*, 49(12), 715–729. September.
76. Zhao, Chengsong, Johanna, Craig, C., H. Earl Petzold, Allan, Dickerman,W., & Eric Beers, P. (2005). The **xylem and phloem transcriptomes from secondary tissues of the arabidopsis root-hypocotyls.** *Plant Physiology*, 138, 803–818.

III. Web page links

77. Aguirre, M. (2002). Las plantas y su estructura. Pp. 10–15 http://www.efn.uncor.edu/dep/biologia/intrbiol/planta1.htm
78. Bartolini, Vitir, S. J., Laghezali, M., & Olmez, H. (2001). Xylem vessel differentiation and microsporogenesis evolution in "Canino" cultivar growing in three different climatic areas: Italy, Morocco, and Turkey. Pp. 87–97. http://virmap.unipi.it/cgi-bin/virmap/vmibo?doc_pubbl:11699404;main;proc
79. Christophertipping, R. (2004). Dispersal adaptations of immature stages of three species of eafhopper (Hemiptera:auchenorryncha:cicallidae). Pp. 23–45. http://www.fcla.edu/FlaEnt/fe87p372.pdf#search='Xilema'
80. Da Costa, A. (2001). As relações hídricas das plantas vasculares. Pp. 23–30 http://www.angelfire.com/ar3/alexcosta0/RelHid/Rhw6.htm
81. Farabee, J. M. (2001). Plant and their structures-I. http://www.emc.maricopa.edu/faculty/farabee/BIOBK/BioBookTOC.html
82. Farabee, J. M. (2001). Plant and their structures-II http://www.emc.maricopa.edu/faculty/farabee/BIOBK/BioBookPLANTANATII.html
83. Farabee, J. M. (2001). Plant hormones, nutrition and transport. http://www.emc.maricopa.edu/faculty/farabee/BIOBK/BioBookPLANTHORM.html
84. Farabee, M. J. (2001). Transport in and out of cells. http://www.emc.maricopa.edu/faculty/farabee/BIOBK/BioBookTOC.html
85. Graus, J., & Pisaneschi, J. (2005). Atlas de anatomía vegetal. Pp. 76–79 http://atlas-veg.ib.usp.br/English/info.html
86. Hölttä, Teemu, (2005). Dynamics of water and solute transport in trees. Pp. 1–28. ethesis.helsinki.fi/julkaisut/mat/fysik/vk/holtta/dynamics.pdf

87. Juairiah L, Nurtjahya, E., Pranitasari, T., & Dorly, J. (2005). Konduktivitas xylem akardan batang tumbuhan pionir di lahan pasca penambangan timah di desa sempan, Bangka. Pp. 67–72. http://www.ns.ui.ac.id/seminar2005/Data/J1B 08.pdf#search=%22%22Xilem%22%22

88. Kimball, J. (2006). Transport of water and minerals in plants. Pp. 12–31 http://users. rcn.com/jkimball.ma.ultranet/BiologyPages/F/FramesVersion.html

89. Koning, Ross E., (2001). Stems. http://149.152.32.5/Plant_Physiology/stemslec.html

90. Martínez, J. (2004). Constraints on water transport posed by xylem embolism: implications for drought and frost resistance in woody plants. Pp. 23–24 http://www.tdx. cesca.es/TDX-0125102–105505/

91. Martínez, M., & Gómez, M. (2006). Vegetative anatomy of two Nymphoides species (Menyanthaceae) Pp. 34–36. http://www.ejournal.unam.mx/biodiversidad/ BIOD7701/BIOD770109.pdf#search=%22%22 Xylem%22%22

92. Pearson Prentice Hall, (2006). Movement of water up xylem vessels. www.phschool. com/.../labbench/lab9/xylem.html

93. Restrepo, H. (2005). Una teoría del movimiento del agua xilemática para interpretar algunas relaciones del clima con la producción de lo cítricos. Pp. 23–24 http://www. agro.unalmed.edu.co/departamentos/iagricola/docs/RIEGOXILpdf#search = % 22 % 22 Xilem%22%22

94. Selles, Ferreira, G. R., Silvva, H., Muñoz, L., Ahumada, R. Y., & Silva, C. (2005). Variación de diámetro de tronco y de baya en Crimson sedles. Pp. 32–3 http://www.inia.cl/platina/investigacion/congresos/docs/2005/2005–11-selles. pdf#search=%22%22Xilem%22%22

95. Sengbusch, Peter, (2003). The Phloem. http://www.biologie.uni-hamburg.de/b-online/e06/06d.htm

96. Torres, J. (2001). Histología vegetal. Pp. 45–47 http://www.inea.uva.es/servicios/ histologia/inicio_real.htm

PART II

EVALUATION OF
MICRO IRRIGATION

CHAPTER 4

EVALUATION OF FILTERS AND EMITTERS UNDER DOMESTIC TREATED SEWAGE WATER

ASHOK R. MHASKE and R. B. BINIWALE

CONTENTS

4.1   INTRODUCTION

Agricultural water demand with conventional resources, using treated waste water represents a viable option [6, 9]. Emitter clogging hazards are major considerations in selecting drip irrigation systems for use with water from open reservoirs, and particularly from that storing secondary treated sewage water. Most clogging factors and agents can be found in these waters [5, 12]. Reservoir water contains a variety of phytoplankton and zooplankton that develop during storage according to the specific conditions prevailing in the reservoir. Suspended particles, which can agglomerate with filaments, slimy or otherwise sticky by-products of microbial activity, are also abundant, especially in earthen reservoirs [14]. There are also aquatic organisms that can grow and proliferate within the pipeline system and, in certain circumstances can develop into a biomass that can clog almost any component of the drip irrigation system [12, 14]. Such problems might be intensified by longer supply lines and slower stream velocities. The nutrients and organic matter brought into the reservoir with wastewater effluent will enhance algae bloom, particularly the slimy blue algae species. Microbial activity in deeper layers increases in reservoirs storing sewage effluent.

Filtration, chemical treatment of the water and flushing of laterals are the means generally applied to control emitter clogging [12]. Changes in procedures and a more careful selection of the system components must be considered for treated sewage water use in drip irrigation. Drip irrigation systems supplied with these treated sewage waters are widely used [8]. Operational difficulties, and even system failures, associated with clogging problems have often been encountered. Filtration is, in fact, mandatory for all drip irrigation systems. A common filtration system consists of primary and secondary filters. Screen or disks filters are used as the secondary downstream safety filters, at the head of the plot's manifold. Usually these filters are cleaned manually. Most manufacturers of drip systems recommend high filtration levels. Consequently, in many cases of low quality waters, the main problems in the operation of drip irrigation systems have been with clogged filters, rather than emitter clogging.

Chlorination to control microbial activity is, apparently, needed when the irrigation water source is a reservoir containing wastewater. In view of the above, an experimental program was initiated in 2013 to test filters and emitters with domestic treated sewage water supplied from tank storing secondary effluent and well water. The trials, carried in field conditions, enabled to closely study the functioning and clogging of various types of emitters and filters and how to alleviate the clogging problems. The aim of this study was to determine the effects of different qualities of domestic treated waste water and well water on clogging of different emitters and filters.

4.2   MATERIALS AND METHODS

4.2.1   EXPERIMENTAL DESIGN

The experiment was conducted under open field conditions in black cotton soil at Agricultural College Campus, Dr. PDKV, Nagpur to study the field performance of different types of emitter and filters under different qualities of water. The filtration units consisted of sand filter followed by disk filter. The study included twelve treatments, which represented the combination between:

A: Two treatments of domestic treated sewage water through phytorid sewage treatment plant at Agriculture college campus and well water (fresh water),

B: Two types of emitters: inline and online, and

C: Three times of operation: 0, 60, and 120 hours.

The measured data included percentage of partial and total clogging of emitters. The collected data were analyzed by using MSTAT program, in a split plot design with three replications.

4.2.2 CHARACTERISTICS OF DOMESTIC TREATED SEWAGE WATER AND WELL WATER

Domestic treated sewage water (DTSW) and well water (WW) were used to test their effects on the performance of the filters and emitters. The domestic treated sewage water was from phytorid sewage treatment plant installed at Agriculture College Campus, Nagpur. The fresh water was from the municipal potable watering system. The treated domestic sewage water (TDSW) and well water (WW) were analyzed in the laboratory for chemical and organic analysis (Table 4.1).

4.2.3 SPECIFICATIONS OF DRIP IRRIGATION SYSTEM

In surface drip irrigation system, two emitters were fitted in and on laterals polyethylene pipe. Table 4.2 shows hydraulic characteristic of two types of emitters. The specifications of sand and disk filter are shown in Table 4.3. The disk filter specifications included 5 cm diameter outlet and inlet, 120 mesh size and maximum flow rate of 30 $m^3.h^{-1}$.

Two different types of filters (sand and disk filters) and two types of emitters (inline and online) were tested. The field layout consisted of a flow meter, two pressure gages installed before and after each filter, and two polyethylene lateral lines per treatment. Lateral line was 45 m long and had 90 emitters connected at a spacing of 0.5 m. The external diameters (O.D.) of the laterals were 20 mm. During each treatment, the system was in operation for about 120 hours with daily operation (4 h) at an operating

TABLE 4.1   Characteristics of Treated Sewage Water and Well Water

Parameter	Sewage water	Treated sewage water	Well water
pH	7.0±0.09	7.1±0.92	7.5±0.84
EC, (dSm^{-1})	0.756	0.602	0.412
SAR	–	0.656	0.615
Carbonates, CO_3, (mg) L^{-1})	–	0.57	0.30
Bicarbonates, HCO_{3-}, (mg) (mg L^{-1})	–	3.81	3.18
Chlorides, Cl$^-$, (mg L^{-1})	–	3.68	1.48
Calcium, (mg L^{-1})	–	5.12	2.68
Magnesium, (mg L^{-1})	–	2.42	0.72
Sodium, Na, (mg L^{-1})	–	1.09	0.80
BOD, (mg.L^{-1})	128	24	12
COD, (mg.L^{-1})	478	60	32
TDS, (mg.L^{-1})	351	399	278
TSS, (mg.L^{-1})	35	19	6
Nitrogen, (mg L^{-1})	3.9	3.70	1.10
Phosphate, (mg L^{-1})	–	1.30	0.26
Potassium, (mg L^{-1})	–	0.32	0.22
Iron, (mg L^{-1})	–	2.85	2.32
Manganese, (mg L^{-1})	–	0.98	0.72
Feacal coli form, cfu per 100 ml	>1100	34	0.0

TABLE 4.2   Hydraulic Characteristic of the Two Emitter Types

Type of emitter	Emitter discharge L.h^{-1}	Emitter discharge exponent	Emitter manufacturing coefficient of variation	Type of emitter discharge
On line Turbo key emitter	3.9	0.4264	4.55	Turbulent
In line	3.97	0.453	4.30	Turbulent

TABLE 4.3   Specifications of Sand and Disk Filters

Specifications	Sand filters	Disk filters
No. of Filters	1	1
Recommended maximum flow rate (m³h⁻¹)	20	30
Maximum operating pressure (kg)	1.0	1.0
Inlet and outlet diameter (inch)	2	2
Length (mm)	900	417
Tank Diameter (mm)	500	25
Wall Thickness (mm)	5.0	3.0
Thickness of media layer (mm)	600	—
Back washing diameter (inch)	2	—
Specification of Media		
Bed area (m²)	0.952	—
Effective diameter of granular media (mm)	1.0 – 1.5	—

pressure of 10 m. On each day of operation and for each type of filter, data observations were the exact time of operation, the total flow volume, clogged emitter's percentage (partially and totally). The number of filter cleaning operations was recorded. Because the head loss in lateral lines was very small, the pressure along the lateral was considered essentially constant. The filters were connected in series.

The filters were cleaned by back flushing, whenever the pressure drop caused by partial clogging of the filter increased to 0.2 kg/cm² [10]. Disk filters were also manually cleaned, by pulling out the filter disks and washing it. All types of filters were manually cleaned and dried at the end of each day of operation. Experimental layout is shown in Figure 4.1.

4.2.4   CRITERIA FOR ASSESSMENT OF EMITTER PERFORMANCE

4.2.4.1   The Reduction of Main Discharge Qreduction Percent

The partial clogging of emitter was calculated as follows [11]:

$$Q_{reduction} = 100\left(\frac{Q_{ini} - Q_{m}}{Q_{ini}}\right) \qquad (1)$$

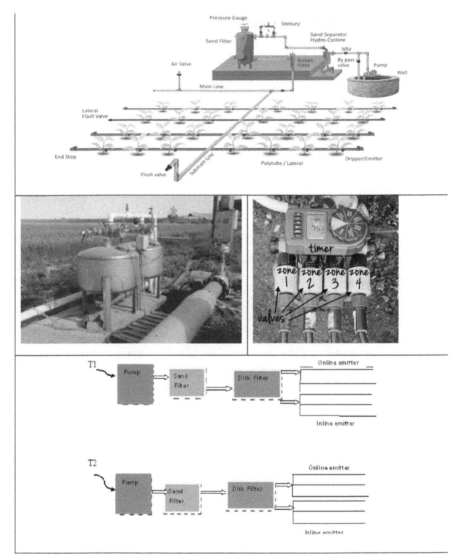

FIGURE 4.1 A typical drip irrigation system (top) with filtration/irrigation controller (center); and layout of the experiment (bottom).

where, $Q_{reduction}$ is reduction of mean discharge in percentage, Q_m is the mean emitter discharge of each lateral of last operation, and Q_{ini} is the corresponding mean discharge of 540 new, unclogged emitter at the same operating pressure.

4.2.4.2   Manufacturer's Coefficient of Emitter Variation (CV)

The manufacturer's coefficient of emitter variation is the measure of the variability of the discharge of random sample of a given make, model and size of a emitter as produced by the manufacturer before any field operation or aging has taken place [2]. The manufacturer's variation is mainly caused by pressure and heat instability during emitter production. The manufacturer's coefficient of variation is determined from the flow rate measurements at different locations and is calculated as follows:

$$CV = [(q_1^2 + q_2^2 + q_3^2 + q_4^2 + \ldots q_n^2 - nq^2)^{1/2}] \div [q\,(n-1)^{1/2}] \qquad (2)$$

where, CV = manufacturer's coefficient of variation; $q_1, q_2, \ldots, q_n$ = discharge of emitter in lph; q = average discharge of the emitter in lph; and n = number of emitter tested.

Classification of CV value according to ASAE standards is shown in Table 4.4 [2]. High CV can occur due to heterogeneous mixture of material used in production of the emitters. Typical values of CV range from 5 to 15%, although higher values are possible [3]. The dripper manufacturing coefficient variation is one of the statistical terms, which can be used to show the drip irrigation system uniformity. Different guidelines have been suggested for Cv, but authors used those recommended by ASAE [2].

4.2.4.3   Field Emission Uniformity Coefficient, EU (%)

At the end of the treatment, the flow from 25 emitters within an irrigation block is collected. Then field emission uniformity coefficient is calculated by Eq. (3) as given by EI Tantawy et al. [7].

$$EU, \% = (q_{min1/4}/q_{ave}) \times 100 \qquad (3)$$

TABLE 4.4   ASAE Recommended Classification of Dripper Manufacture Coefficient of Variation [2]

Classification	Excellent	Average	Marginal	Poor	Unacceptable
Cv (%)	< 5.0	5.0–7.0	7.0–11.0	11.0–15.0	> 15.0

where: EU = the emission uniformity (%); $q_{min1/4}$ = measured mean discharge of the lowest quartile in lph; and q_{ave} = measured mean discharge of all emitter in lph.

4.2.4.4   Dripper Exponent (x)

The emitters are the most important part of the drip irrigation. Emitters with high degree of pressure compensation (x = 0) are technically possible, although the ideal emitter has not yet been designed. The emitters flow rate may fluctuate due to pressure variation along the dripper line, and accidental restriction resulting in non-uniform water application [4]. Emitters discharge rate is a function of operating pressure that is described by a power law.

$$q = k \, [H]^x \tag{4}$$

where, q = emitter discharge rate in lph; k = emitter constant, including a conversion factor to make units uniform on both sides of equation; H = operating pressure in m; and x = emitter exponent. The constants k and x define characteristics of an emitter.

For fully laminar flow regime, emitter must be very sensitive to pressure variation and the value of x must be 1.0. This means that the pressure variation of 20% may result in ± 20% emitter flow rate variation. Most non-compensating emitters are always fully turbulent with x = 0.5, indicating the pressure variation of approximately 10%. On other hand for compensating emitter, pressure variation cause very small variation in discharge. For compensating emitter, the x ranges from 0.1 to 0.4. An ideal pressure compensating emitter would have an x = 0 [3, 4]. Equation (4) was used to calculate x values in this study.

4.2.4.5   Emitter Flow Variation (qvar)

Emitter flow rate variation q_{var} is given by following equation [2, 10].

$$q_{var} = 100 \times [q_{max} - q_{min}] \div [q_{max}] \tag{5}$$

where, q_{var} = variation of the average flow rate (%); q_{max} = maximum flow rate at pressure of 1.0 kg/cm^2 at the same water temperature along the line; q_{min} = minimum flow rate at pressure of 1.0 kg/cm^2 at the same water temperature along the line. Emitter flow rate variation less than 10% is considered desirable.

4.3 RESULTS AND DISCUSSION

4.3.1 WATER QUALITY

The data pertaining to the characteristics of domestic treated sewage water and well water used for irrigation were analyzed and are reported in Table 4.1. The characteristics of irrigation water (domestic treated sewage water and well water) was assessed for its suitability for irrigation with respect to pH, EC, SAR, carbonate, bicarbonate, calcium, magnesium, sodium, BOD, COD, nitrogen, phosphate, TDS and potassium [1]. The domestic treated sewage water as well as the well water was slightly alkaline in reaction.

The pH of the treated sewage was 7.1 slightly lower than the well water, whereas EC of 0.602 dS.m^{-1} was found higher than well water (EC = 0.412 dS.m^{-1}). Calcium was the dominant cation followed by magnesium and sodium, although the sodium content was slightly higher in treated sewage water. The sodium adsorption ratio (SAR) of treated sewage and well water was observed less than the critical limits. Carbonate and bicarbonate contents in the treated sewage water were observed slightly higher than in well water, whereas chloride content was observed 2 to 3 folds higher in sewage water than in well water. SAR of treated and well water was noticed minimum, which indicated that both waters were suitable for irrigation. Looking to the concentration of BOD and COD, the raw sewage water was rated as unsuitable for irrigation purpose when compared with the prescribed limit of 100 and 250 mg.L^{-1} for BOD and COD, respectively. In treated sewage water, the higher contents of NPK were observed. In treated sewage water, N and P was observed 3 and 5 times higher, respectively. However, the K content was noticed slightly higher than the well water [13].

4.3.2   MEAN DISCHARGE REDUCTION

Emitter discharge generally decreases from the beginning to the end of lateral due to pressure loss in the lateral [15]. The difference in mean emitter discharge between first ¼ emitters and last ¼ emitter was 0.3% for the first measurement using fresh water. The emitter discharge in DTSW treatment was lower than that of WW treatment. The reduction of discharge was more in inline emitter than online emitter.

4.3.3   EFFECTS OF DTSW AND WW TREATMENTS AND EMITTER TYPE ON PARTIAL AND TOTAL CLOGGING OF EMITTERS

In case of total emitter clogging percentage, the differences between all the treatments [DTSW and WW, emitters type (Inline and Online) and operation time (0, 60, 120 h)] were significantly different except the combination difference between water types in their effects on partial emitter clogging and total emitter clogging percentage. The partial clogging and total clogging percentage were significantly different due to different concentrations of organic matter and chemical in DTSW and WW. The results revealed that the on line emitter were better than inline emitter under different treatments. The data in the Table 4.5 revealed that low clogged emitters were in T_2 and high clogged emitters were in T_1. The CV for DTSW treatment is similar to that of the for the well water treatment. The CV was increased with operational time reaching a value of 7.98% at the end of the experiment [11]. The effect of the source of the water on emitter is presented in Table 4.5. It is evident that with partial clogging levels, WW at 120 hours was reached with DTSW in 60 hours itself. However, in case of total clogging, level with DTSW was higher at 60 hours than that of WW at 120 hours. This indicates that if DTSW is used then emitters have to be cleaned more frequently.

4.3.4   EFFECTS OF DTSW AND WW TREATMENTS ON PARTIAL FILTER CLOGGING PERCENTAGE

The effects of different DTSW and WW on partial clogging percentage in sand and disk filters are shown in Figure 4.2. The partial clogging

TABLE 4.5 The Effects of DTSW and WW Treatments, and Emitter Types on Partial and Total Clogging of Emitters

Treatment	Emitter type	Partial clogging				Total clogging			
		Time of operation (h)			Mean	Time of operation (h)			Mean
		0	60	120		0	60	120	
T1	In line	3.97	6.95	15.54	8.82	2.0	5.61	7.37	4.99
	On line	3.00	6.06	13.44	7.50	1.26	4.45	5.58	3.76
	Mean	3.49	6.51	14.49	8.16	1.63	5.03	6.48	4.38
T2	In line	1.75	3.01	6.84	3.87	0.88	2.46	3.24	2.19
	On line	1.32	2.76	6.08	3.39	0.67	2.33	2.78	1.93
	Mean	1.54	2.89	6.46	3.63	0.76	2.40	3.01	2.06
Mean	In line	2.86	4.98	11.19	6.35	1.44	4.04	5.31	3.59
	On line	2.16	4.41	9.76	5.45	0.97	3.39	4.18	2.85
	Mean	2.51	4.70	10.48	5.90	1.21	3.72	4.75	3.22
L.S.D at 0.05									
	A		NS			NS			
	B		0.23			0.15			
	C		0.28			0.20			
	AxB		0.32			0.22			
	AxC		0.40			0.27			
	BxC		0.40			0.27			

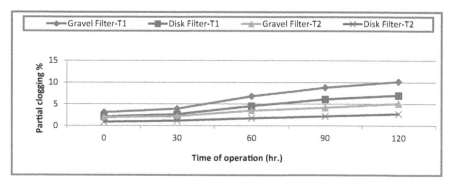

FIGURE 4.2 Effects of DTSW and WW on partial filter clogging percentage.

percentage of sand and disk filters ranged from 3.12 to 10.15% and 2.10 to 7.12%, respectively in T_1. Similarly in T_2, partial clogging was 2.18 to 4.98% in sand filter and 1.3 to 3.98% in disk filter. The partial clogging percentage in sand and disk filter in T_1 was higher that may be because of increasing the organic matter and chemical contents in the DTSW than WW [7]. This indicates that use of DTSW may clog the filters earlier than the WW. The partial clogging percentages of sand filter in T_1 is 5.17% more than that in T_2. In disk filter, it is 3.94% after 120 hours of operation.

4.3.5   MEAN DISCHARGE REDUCTION PERCENTAGE FOR DIFFERENT TYPES OF FILTERS

Percentage of reduction of mean discharge for different types of filters (sand and disk filters) at the start and end of the experiment during 120 hours of operation of 120 is shown in Figure 4.3. The reduction of mean discharge percentage of sand filter ranged from 4.65 to 12.56% in T_1 compared to 2.94 to 7.69% in T_2. The reduction of mean discharge percentage in T_1 is more due to increasing of organic matter and chemical content in the DTSW. The increasing discharge reduction of filters increased the filtration efficiency due to preventing organic matter and nonorganic sedimentations in the DTSW and WW and effect on the clogged emitters. The clogged emitters range is not sufficient to account for the reduction in mean discharge so there must be the problems of partial clogging as well. The discharge reduction

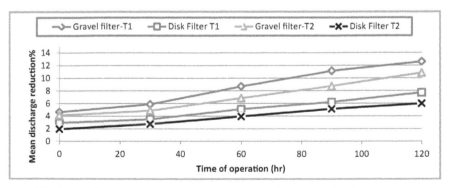

FIGURE 4.3   Effects of DTSW and WW on mean discharge reduction percentage for different types of filters.

percentage of sand filter in T_1 is more than T_2 by 14.01%. Similarly in disk filter, the discharge reduction percentage in T_1 is more than T_2 by 22.36% after 120 hours of operation.

4.3.6 FIELD EMISSION UNIFORMITY COEFFICIENT

Field Emission uniformity coefficient (*EU,* %) measured at the end of two irrigation water treatments is shown in Figure 4.4. The EU (%) classification shows: 90.0% excellent, 80.0–90.0% good, 70.0–80.0% fair, 60.0–70.0% poor, and <60.0% unacceptable [2]. Emission uniformity coefficient values equal to zero mean that at least one quarter of the dripper tested were completely clogged. EU (%) was generally better in T_2 than treatment T_1 as expected in both inline and online drippers; online emitter gave better EU (%) than inline dripper. Online T_2 was good for 120 hours of operation. The performance of the emitter with TSW was in range of good upto 60 hours and then fair upto 120 hours, which is similar to partial clogging of the emitters. The results showed that increasing the total suspended solids, BOD in T_1 more than T_2 led to decreasing of EU percentage in T_1 as compared to T_2.

4.3.7 INFLUENCE OF THE DTSW AND WW CHARACTERISTIC ON PERFORMANCE OF EMITTERS AND FILTERS

The inline emitters with a similar discharge were more sensitive to clogging compared to online emitter. The results showed that there was more

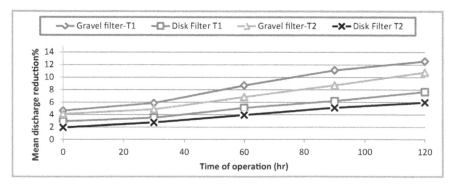

FIGURE 4.4   Field Emission uniformity coefficient (EU) of drippers.

influence of the domestic treated sewage water on the performance of the drip irrigation systems, for the same type of emitter and filter as compared to the well water. When the total suspended solids (TSS) and organic matter were increased, the percentage of totally clogged emitters was also increased, whereas the mean emitted discharge, emission uniformity coefficient, and the operating time of the filters between cleaning operation were decreased. The highest mean manual operation and lowest back flushing operation were recorded in disk filters.

4.4 CONCLUSIONS

This chapter discusses effects of DTSW and WW characteristics on the performance of the drip irrigation systems for different types of the emitters and filters. With increase in total suspended solids and organic matter content, the percentage of total clogged emitters was also increased, whereas the mean discharge of emitters, emission uniformity coefficient, and the operating time of the filters between cleaning operation were decreased.

The inline emitters with a similar discharge were more sensitive to clogging compared to online emitters. Gravel media filter decreased the emitter clogging and increased the performance. The disk filter of good quality is cheaper and simpler to manage and give good performance similar to sand filter, provided there should not be algae present in the water. The results showed that although DTSW can be used through drip system, it requires cleaning at half intervals than well water. Thus there is a need to develop drippers compatible to such water quality, because treated water is costly and has to be used with efficient irrigation methods. Automatic back flushing systems are also preferable to avoid the contact between wastewater and irrigator with short operating times of the filters between the manual cleaning operations. Increasing the total suspended solids and organic matter, BOD and Ca^{++}, Mg^{++} in T_1 more than T_2 led to decrease in clogging percentage.

4.5 SUMMARY

The experiment was carried out under open field condition in black cotton soil at Agricultural College Farm in Nagpur to study the performance of filters and emitters under domestic treated sewage water treated through phytorid sewage treatment plant and well water during 2013. The trials were conducted with two water supplies: domestic treated sewage water (DTSW, T_1) and well water (WW, T_2). The measured data included partial and total clogging emitter percentage. The collected data were analyzed by using MSTAT program, in a randomized block design.

The results show a significant difference due to increasing concentration of wastewater organic materials (BOD and TSS) and Calcium, Iron and Magnesium contents. The sand media filter followed by the disk filter gave better performance in T_2 than T_1 after 120 hours of operation, respectively. The partial clogging percentages of sand filter in T_1 was more than in T_2 by ratios 5.02% and in disk filter 3.14%; and the mean discharge reduction percentages 1.76% in sand filter and 1.72% in disk filter. Emitter's emission uniformity percentage in T_2 was better than T_1. The highest mean manual operation and lowest back flushing operation were recorded in disk filters.

KEYWORDS

- air pollution
- ASAE
- BOD
- calcium
- domestic treated sewage water
- drip irrigation
- emission uniformity
- emitters
- filters
- iron
- magnesium
- micro irrigation
- organic material
- percentage clogging

- phytorid
- treated sewage effluent
- TSS
- wastewater
- well water

REFERENCES

1. Al-Jasser, A. O. (2009). Grey water characterization in Saudi Arabia. Abstract Book 15th by International Union of Air Pollution Prevention and Environmental Protection Association (IUAPPA), World Clean Air Congress: Achieving environmental sustainability in a resource hungry World, p. 20–21.
2. ASAE Standards, (1996). *Amer. Soc. of Agric. Eng.,* St Joseph, MI: 792–797.
3. Boswell, M. J. (1985). Design characteristic of line source drip tubes. *Proc. of the 3rd Int. Drip/ Trickle Irr. Cong.,* Vol. 1, California, USA. p. 306–312.
4. Braud, H. J., & Soon, A. M. (1980). Trickle irrigation design for improved application uniformity. ASAE and CSAE Mtg. paper no. 79–2571, Winnipeg, Canada.
5. Bucks, I. V. Nakayama, F. S., & Gilbert, I. G. (1979). Trickle irrigation water quality and preventative maintenance. *Agric. Water Manage.,* 2(2), 149–162.
6. Capra, A., & Scicolone, B. (1998). Water quality and distribution uniformity in drip/ trickle irrigation systems. *J. Agric. Eng. Res.,* 70, 355–365.
7. EI-Tantawy, M. T., Matter, M. A., & Arafa, Y. E. (2009). Filters and emitters performance under treated waste water. *Misr. J. Ag. Eng.,* 26(2), 886–904.
8. Feigin, A., Ravina, I., & Shalhevet, J. (1991). *Irrigation with treated sewage effluent. Management for Environmental Protection.* Springer, Berlin, Heidelberg, New York.
9. Goyal, Megh, R. (2015). *Research Advances in Sustainable Micro Irrigation, Volumes 1 to 10.* Oakville, Canada: Apple Academic Press Inc.
10. Keller, J., & Bliesner, R. D. (1991). *Sprinkler and Trickle Irrigation.* AVI Book, New York, USA. p. 652.
11. Liu, H., & Huang, G. (2009). Laboratory experiment on drip emitter clogging with fresh water and treated sewage effluent. *Agric. Water Manage.,* 96, 245–756.
12. Nakayama, F. S., & Bucks, D. A. (1991). *Water Quality in Drip/Trickle Irrigation.* A VI Book. New York. USA. p. 187–102.
13. Oswald, W. J. (1989). Use of wastewater effluent in agriculture. *Desalination,* 72, 67–80.
14. Ravina, I., Paz, E., Sofer, Z., Marcu, A., Shisha, A., & Sagi, G. (1992). Control of emitter clogging in drip irrigation with reclaimed waste water. *Irrig. Sci.,* 13, 129–139.
15. Ravina, I., Paz, E., Sofer, Z., Marcu, A., Schischa, A., Sagi, G., Yechialy, Z., & Lev, Y. (1997). Control of emitter clogging in drip irrigation with stored treated municipal sewage. *Agric. Water Manage.,* 33(2), 127–137.

CHAPTER 5

EVALUATION OF DRIP IRRIGATION SYSTEMS FOR WATER MANAGEMENT IN ORCHARDS

RICHARD KOECH, ISA YUNUSA, IAN NUBERG, and
DAVID PEZZANITI

CONTENTS

5.1   INTRODUCTION

Agriculture is by far the largest consumer of fresh water, contributing to approximately 70% of all withdrawals annually on a global scale. However, with the increasing human population, it is facing competition for the scarce water resources from domestic and industrial users. This has led to the promotion of irrigation methods, particularly the drip system, that are considered to be more water and labor efficient [10].

The performance of a drip irrigation system, commonly evaluated in terms of the uniformity of water application, is largely influenced by the hydraulic design of the system, the manufacturer's preciseness in the production of the drippers and emitter spacings [10, 15]. The significance of the influence of the manufacturer's preciseness is attributed to the fact that emitter flow paths are small (typically less than 2 mm in diameter), and therefore a small deviation in diameter would result in relatively large deviations in discharge [12]. As opposed to the sprinkler method and other systems of irrigation where water is spread over the entire surface, drippers supply water directly to small areas in the vicinity of the plant roots at low flow rates (typically 0.5–20 lph). Therefore drippers or emitters, and the uniformity with which they supply water are of vital components of any drip irrigation system design.

Drip systems may be evaluated under laboratory conditions or *in situ*. Tests that require a higher degree of precision, for instance the assessment of the manufacturer's variation in the flow rate of a batch of drippers, are ideally undertaken in a laboratory where environmental conditions can easily be controlled. On the other hand, the efficiency of a drip irrigation design is normally evaluated on site. This chapter covers both the laboratory and field evaluations of drip systems using data collected in Australia and identifies criteria for their improvement.

Full potential of any irrigation technology is achievable when matched with on-farm water management regimes. The capacity of drip system to supply water directly close to a restricted target areas (root zone) makes it particularly suited for orchard irrigation allowing water supply to individual tree/vine system be effectively managed. Thus the common technique of crop factor (K_c), which is an empirical parameter and often applied to the whole farm would not generally be appropriate for orchards where trees and/or vine are grown. For such systems, water management with drip irrigation systems should be based on temporal water for demand to meet potential transpiration, rather than evapotranspiration that includes losses through soil evaporation from exposed soil surfaces. Such an approach would also subsume considerations soil factors, such as texture, structure and hydraulic characteristics. Simple approaches are needed that improve on the commonly used K_c concept.

This chapter deals with both the field and laboratory performance evaluation of drips systems with the primary goal of assessing the effectiveness of the technology in terms of optimizing water-use efficiency and minimizing water losses in irrigated systems. The field design of the system, including irrigation scheduling approaches for efficient water application is also discussed.

5.2 REVIEW OF HYDRAULICS OF DRIP IRRIGATION

Emitters or dippers are generally classified according to their hydraulic characteristics. Traditionally, they are categorized according to how they are attached to the lateral, discharge characteristics, and their design and construction [6] (Table 5.1). The relationship between flow rate, q, and inlet pressure in the emitter, p, can be characterized by the generalized orifice equation of the form:

$$q = k \times p^{m} \tag{1}$$

where, k is a constant and m is the emitting discharge exponent.

Using the values of flow rates q and their corresponding inlet pressure p, the ISO 9261 [11] recommends the use of the least square method to determine the coefficient k and exponent m:

$$m = \frac{\sum \lg p_i (\lg \bar{q}_i) - \frac{1}{n}(\sum \lg p_i)(\sum \lg \bar{q}_i)}{\sum (\lg p_i)^2 - \frac{1}{n}(\sum (\lg p_i))^2} \tag{2}$$

where, i is 1, 2, 3, ... n; n is the number of pressure values; $\bar{q}$ is the mean flow rate; and p_i is the emitter inlet pressure. Equation (3) is used to determine the constant k:

TABLE 5.1   Categories of Drippers*

Classification	Description
Attachment to the lateral	Online
	Emitters have to be installed into the drip line or lateral
	Inline
	Emitters are embedded or integrated into the drip line during manufacturing
Discharge characteristics	Pressure compensating (PC)
	Designed to discharge at constant flow rates over a wide range of operating pressure
	Non-pressure compensating (nPC)
	Flow rates vary according to pressure
Design and construction	Long-path emitters
	Emitters with smooth and long flow paths
	Tortuous-path emitters
	Emitters with relatively long flow paths characterized by sharp bends, contractions, expansions and wall friction
	Short-path emitters
	Emitters with short flow paths
	Orifice emitters
	Water exits through a small-diameter openings or series of openings
	Vortex emitters
	The discharge has a circular pattern due to the design of the flow path

*Adapted from Dutta [6].

$$k = exp\left(\left\langle \frac{\sum In\,\overline{q}}{n} - \frac{m*(\sum In\,p_i)}{n} \right\rangle\right) \tag{3}$$

where all the variables are as defined above. Alternatively, k and m may be determined directly using the Excel regression function. It is obvious from Eq. (1) that the higher the value of exponent m, the more will the flow rate q be affected by pressure, and vice versa. The value of the exponent m for pressure and non-pressure compensating (nPC) orifice and nozzle emitters is approximately 0 and 0.5, respectively [12]. However, using a power function to characterize the flow rate-pressure profiles for pressure compensating (PC) drippers may be inappropriate; a linear relationship is ideally expected and hence a linear function is most applicable. The constant k is a function of the size of the opening and its characteristics.

Both hydraulic and agronomic factors are considered in the design of drip irrigation systems. The operating pressure, frictional losses, size and length of the lateral and water temperature are the key hydraulics factors that will influence the design of the system. On the other hand, agronomic factors considered include the soil infiltration characteristics and the plants to be grown.

5.2.1 UNIFORMITY OF WATER APPLICATION IN DRIP SYSTEMS

Uniformity of water application in irrigation is a critical performance criterion for drip systems since it determines the evenness of the distribution of water across the targeted area. A high uniformity reflects a fairly even application, whereas a low uniformity is characterized by portions of area having significantly higher or lower than the average application. This may mean that crop growth may be hampered either by water logging or insufficient water. Therefore one of the goals of a good irrigation design is a water distribution pattern that that achieves an acceptable uniformity threshold.

The procedure used to determine the uniformity of water application in drip systems involves the use of catch-cans or buckets placed under each dripper to determine the flow rate. This methodology is described in greater detail in the next section of this chapter.

The ISO 9261 [11] and ASAE [3] recommend the use of the coefficient of variation, c_v, to determine the uniformity of flow or emission rates of a sample of drippers. This requires the determination of average flow rate ($\bar{q}$) and the standard deviation (s_q) of the sample. The c_v (%) of the sample is as follows:

$$c_v = \frac{s_q}{\bar{q}} * 100 \tag{4}$$

Equation (4) is normally used in laboratory testing to determine the manufacturing variability of flow rate of a random sample of a dripper model before they are used in the field. The distribution uniformity (DU) is commonly used to determine uniformity of water application *in situ* [14]. DU emphasizes the under-watered portions of the field and may be expressed as:

$$DU = \frac{Mean\,of\,the\,lowest\,25\%\,of\,applied\,depths}{Mean\,applied\,depths} \qquad (5)$$

5.3 EVALUATION PROCEDURES

Understanding the hydraulic characteristics of the drip system is important in improving the water-use efficiency, reduction of water losses and overall success of the system. The relationship between the operating pressure and the discharge of the emitter, and the uniformity of water application are the two most commonly used parameters to characterize the drip system. As mentioned above, this can be undertaken under controlled conditions, for instance in a laboratory, or in the field.

5.3.1 LABORATORY METHOD

Laboratory approach often adopt the international standard, ISO 9261 [11], which provides guidelines on emitter test specimens and conditions, as well as test methods and requirements. The standard specifies that tests are conducted at ambient and water temperature of 23°C ± 3°C using a sample of 25 emitters. Three categories of tests are undertaken: (i) uniformity of flow rate; (ii) flow rate as a function of inlet pressure; and (iii) determination of emitter/emitting unit exponent. We implemented this procedure recently at the Australian Irrigation and Hydraulics Technology Facility (AIHTF), University of Australia, and will now be described. The tests were undertaken using: (i) online nPC drippers with nominal flow rates of 2, 4 and 8 L/h and recommended operating pressure of 100 kPa, and (ii) online PC drippers with nominal flow rate of 4 L/h and recommended operating pressure of 100–400 kPa.

The test rig consisted of a water reservoir, a pump, 40 mm manifold and short lengths of 12 mm pipe attached to either sides of the manifold.

A close mesh metal bar grating was used to cover the reservoir, which also acts as a platform on which the one-liter plastic catch-cans used to collect water from each emitter were placed. The drippers, spaced at 30 cm, were attached to the 12 mm nominal diameter pipes.

The system was left running for approximately one hour before commencing the tests in order to expel any air that may have been trapped in the pipe network or the drippers. Testing was undertaken by sequentially (after about 5 seconds) placing catch-cans under the emitters while the pump was running. The same order was maintained when removing the catch-cans from the test rig at the end of the test. The test duration was approximately 5 minutes, measured using a stop watch. The test pressures ranged from 80–240 kPa, with increments of 20 kPa, leading to 9 pressure levels or tests. The inlet pressure, water temperature and salinity, and ambient temperature were measured for each test. To estimate the mass of water lost due to evaporation, a catch-can half full of water was placed in the vicinity of the test rig. The difference in the mass of water before and after each test was assumed to be evaporative loss, and was consequently added to the mass of water collected in each catch-can, determined by weighing. The flow rate, Q (m³/s), of each emitter was determined as follows:

$$Q = \frac{m}{\rho t} \tag{6}$$

where m (kg) is the mass of water in each catch-can, ρ (kg/m³) is the density of water and t (seconds) is the test duration. The density of water in Eq. (6) was corrected for temperature, pressure and salinity, which are the main factors affecting density of water.

The relationship between the flow rate and pressure was determined by plotting the data in Excel, and determining the exponent m and coefficient k using Eqs. (2) and (3) respectively, while the c_v was calculated using Eq. (4).

5.3.2 FIELD METHOD

The general principles used in the laboratory evaluation of drippers described above are also applicable in the field assessment of the drip system. However, the DU (Eq. 5) rather than the c_v (Eq. 4), is normally used to characterize the distribution of water application. This implies that the water collected in catch-cans will be converted into depths.

An example of a field evaluation procedure for a drip system is described in WATERWISE [14]. The procedure includes two key steps that are not applicable to tests undertaken in a laboratory: (i) the need to select a representative sample of drippers; and (ii) to measure area wetted by drippers. This is done by selecting four dripper laterals along an operating sub-main; one each near the inlet and the outer end of the sub-main, and two in between. Four drippers are then selected from each lateral for evaluation, one each must be near inlet and end of the lateral. The selection of the drippers is meant to account for the variation of the dripper hydraulics characteristics according to the length of the lateral and the sub-main. To estimate the area wetted by each dripper, it is necessary to dig out the soil directly beneath the dripper to about 15–30 cm.

5.3.3 QUANTIFYING UNCERTAINTY IN MEASUREMENTS

Uncertainty is a parameter that characterizes the dispersion of values caused by random and systematic errors associated with the measurement. Practically, this means that the true value of a measurement differs from its measured value by the magnitude of the associated measurement uncertainty. The procedure commonly used by metrological facilities to determine uncertainties in measurements is described in Bentley [4], an ISO guide resource book to uncertainty in measurement [11]. According to this standard, the true measurement of the flow rate (Q) from a dripper may be expressed as:

$$Q = Q_m + U \tag{7}$$

where Q_m is the measured flow rate and U is the uncertainty associated with the measured flow rate. The sources of errors or uncertainty components may be quantified using statistical methods, for instance variances or standard deviations. An example of an uncertainty assessment of the variation coefficient of drip irrigation emitter flow rate is illustrated by Zhao et al. [20]. It is clear from Eq. (4) that the magnitude of the uncertainty is an indication of the quality of the measurement. A low uncertainty signifies a better control of the measurement process or a higher accuracy of measurement, and vice versa.

An important step in the calculation of measurement uncertainty is the identification of all potential sources of error, both random and systematic. Some of these error sources relate to the equipment used, while others may be caused by the actual procedure used during testing. In the dripper tests undertaken at the Australian Irrigation and Hydraulics Technology Facility (AIHTF), the following were identified as uncertainty components: calibration, drift and resolution of the measuring scale, correction of water density for uncertainty in temperature and salinity, stop watch resolution, test time error, uncertainty in evaporation loss, correction for buoyancy (air displaced from collector) and uncertainty due to error in pressure.

5.4 RESULTS AND DISCUSSION

The results and the key descriptive statistics obtained from this laboratory case study are presented in Figures 5.1–5.4. The results are discussed below in three subsections: uniformity of flow rate; flow rate as a function of inlet pressure; and determination of emitter/emitting unit exponent.

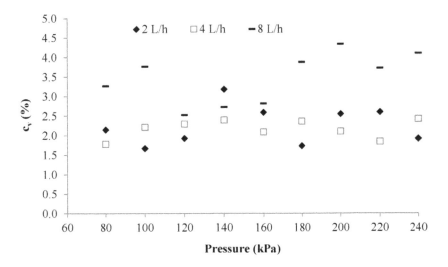

FIGURE 5.1 Relationship between c_v and pressure (nPC drippers).

5.4.1   UNIFORMITY OF FLOW RATE

The c_v (Eq. 4) of the flow rate of the of the dripper samples are shown in Figures 5.1 and 5.2 for the nPC and the PC drippers, respectively. The c_v in this case reflects the manufacturer's variation in the flow rate of the dripper samples tested since the drippers were unused. The maximum c_v was found to be 2.59, 2.41, and 4.34% for the 2, 4 and 8 L/h dripper model respectively (Figure 5.1). There was no distinct pattern of c_v with the operating pressure for the 2 and 4 L/h model. The c_v of the 8 L/h model generally seemed to increase with pressure. It is clear from Figure 5.1 that the c_v of the 8 L/h model are generally higher than that of the other two models in the nPC category. It is also worth noting that the uncertainty of measurement increases with the dripper nominal flow rate rating. This is because the test duration for the 8 L/h model was lower than the other two models, and as indicated earlier, the test time is a potential source of error in measurement. The shorter the test duration (and therefore less volume collected in the catch-can), the higher the uncertainty in measurement, and vice versa.

The pattern of c_v and the pressure for the PC dripper is shown in Figure 5.2. The pressure range covered (100–400 kPa) is the manufacturer's

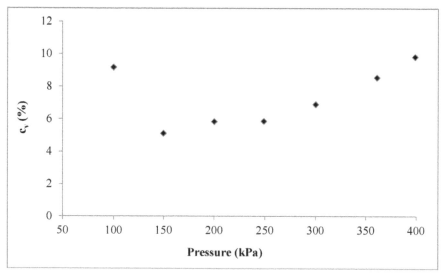

FIGURE 5.2   Relationship between c_v and pressure (PC dripper).

recommended operating pressure for this particular dripper model. The is c_v is shown in this figure to be higher both in the minimum and the maximum operating pressure level. There is a general increase in c_v with pressure from the 150 kPa pressure level. The nPC and PC drippers were of different models and therefore the results cannot be compared directly. Suffice to say that the c_v of PC (with a nominal flow rate of 4 L/h) was generally higher than corresponding nPC of the same flow rate rating.

As indicated earlier, non-uniform irrigation water application may mean that plant growth will be adversely affected. An analysis of the manufacturer's c_v of drippers (Figures 5.1 and 5.2) is therefore important for the design of a drip system. The field uniformity of water application of drippers is likely to be lower because on non-uniform terrain and emitter clogging, among other factors.

5.4.2   FLOW RATE-PRESSURE PROFILES

The profiles of flow rate and nominal pressure for the nPC and the PC drippers are shown in Figures 5.3 and 5.4, respectively. For each pressure level, the average flow rates of the 25 emitters were used to plot the graphs. Whiskers have been used in these figures to depict the standard deviations in the flow rates.

For the nPC, whose flow rate is influenced by the operating pressure, the data best fitted a power function with R^2 of 0.99 or better (Figure 5.3). For all the three dripper models, the emitter exponent (m) was approximately 0.5, which is consistent with results presented in literature [12]. On the hand, the coefficient k, which is a function of the emitter size, as expected increased with the flow rate rating of the emitters. The regression functions shown in Figure 5.3 may be used to design a drip system.

One advantage of nPC is the possibility of increasing the flow rate in instances where the pressure in the drip system can be increased, for instance using a variable speed water pump. In this case the regression equations (Figure 5.3) will be valuable in the design of the system.

Figure 5.4 shows that at the average emitter discharge was highest and lowest at the operating pressure of 200 and 400 kPa respectively. The figure also shows that the average flow rate measured was slightly lower than the nominal flow rate specified by the manufacturer for this

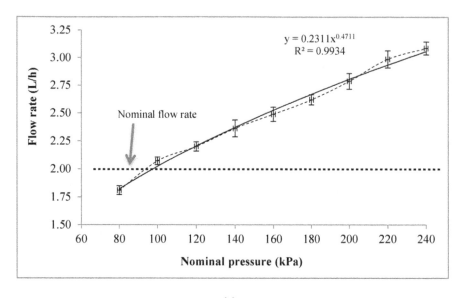

(a)

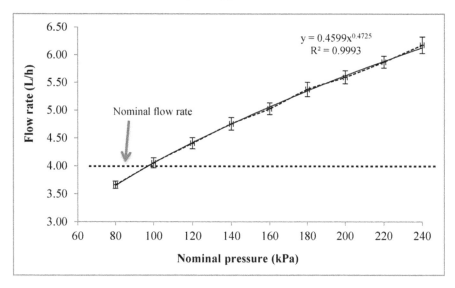

(b)

FIGURE 5.3 Continued

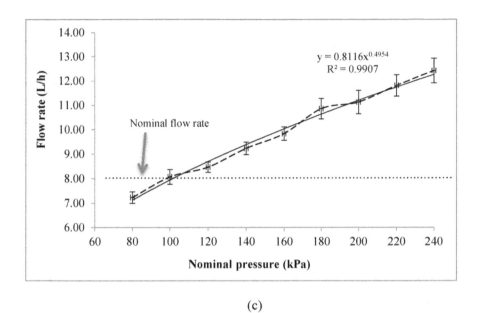

(c)

FIGURE 5.3   Distribution of flow rates according to operating pressure (nPC emitters).
(a) Model 2 L/h; (b) Model 4 L/h; (c) Model 8 L/h.

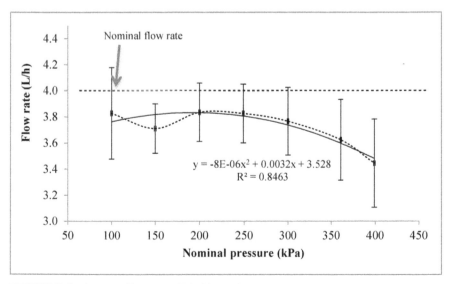

FIGURE 5.4   Average flow rates (PC drippers).

sprinkler model. The polynomial function best suited the data, with R^2 of 0.8463. Ideally, the discharge [rate] estimated by these models is supposed to be independent of pressure. The PC drippers are ideal for use in areas with an uneven terrain, and in cases whereby plant nutrients are distributed through the drip system.

Deriving the full benefits in water-use efficiency from drip systems when deployed to field requires an ability to match water delivery to transient requirements for water by crop for transpiration. This is generally approached based on soil, crop and/or environmental parameters, and will be discussed in the following sections.

5.4.3   EFFICIENT WATER APPLICATION WITH DRIP SYSTEMS

By confining water application to a confined space, drip irrigation systems potentially minimize water wastage and should be expected to produce high water-use efficiency for irrigated crops. Drip irrigation systems are quite ideal for row crops, such as trees and vines since the drippers the distribution of the drippers can be aligned with the discrete arrangement of individual plants in orchards. This is unlike broadacre cropping systems in which high planting rates often results in complete canopy cover at some stage during the season. Profitable and sustainable irrigation practices require three key groups of factors:

1. Water requirement by the crop at any given period during the season. This depends on the type of crop grown, its morphology (canopy and rooting characteristics), phenology (phasic development and length of growing season), and its sensitivity to soil and ambient characteristics.
2. Evaporative demand of the location. This sums the impact of local meteorological variables such as ambient weather conditions, mainly solar radiation, temperature, wind, humidity, and exposure, which exert water demand on the local vegetation.
3. Soil hydrologic characteristics that determine distribution and storage of water, and its subsequent availability to the crop. Properties such as texture, structure, depth, infiltration and internal drainage have direct influence on spatial and temporal availability

of soil water. The topography of the landscape would also have an impact on soil-water availability since it affects the depth of the soil, hence its water storage capacity, and the potential for runoff from rainfall and/or irrigation.

Each of these three factors needs to be considered in designing efficient on-farm irrigation practices. This however, not easily attainable since each of the factors has its own complex internal dynamics that are not easily characterized and integrated, and to be readily applied for on-farm irrigation water management. For practical purposes, therefore, on-farm water management relies on satisfying crop water requirements for optimum evapotranspiration (ET) under prevailing management and meteorological conditions. This concept groups all the key drivers to just two factors that are relatively easier to quantify, viz. (i) temporal crop water-use and (ii) temporal potential water demand. Temporal water-use sums the influence of management, weather and immutable soil characteristics; whereas the potential water demand sets the upper limit for water-use/requirement needed to support the crop if all other factors are optimized. These two components are integrated into one single and simple concept of crop factor (K_c):

$$K_c = \frac{ET}{E_o} \tag{8}$$

where, the E_o is potential evaporation that integrates the influence of temperature, humidity, rain fall, solar radiation. The concept derives from the understanding that the water a crop uses or requires is dependent on the local weather conditions. Although the K_c concept seems to be simple, it presents several critical challenges in terms of precisely determining the E_o and also the ET. Historically E_o was obtained from pan evaporation, but pan evaporimeters have since become anachronistic due to uncertainties arising from their location, correcting for rainfall, and difficulty in estimating rates at fine (hourly) time scales. Nowadays pan evaporimeters have been replaced with micrometeorological techniques to calculate reference evapotranspiration (ET_o) based on Penman-Monteith equation [2]. This replaces the E_o in Eq. (7) with ET_o.

The K_c concept as given in Eq. 7 is empirical in nature since it requires some foreknowledge of optimum ET with the crop expressing minimal constraints. This may not always be possible especially with new crops and in new environments. Under such circumstance, reliable estimates of ET need to be undertaken. This used to rely on growing crops on lysimeters and calculating the amount of water used [1, 17, 18], but the technique is not easily replicated due to cost. Micrometeorological approaches such as the Penman-Monteith and even energy balance systems, for example, the Bowen ration and eddy covariance, require quantifying canopy conductance to quantify water use by plants as distinct from combined soil and plant surfaces.

An overarching objective of the K_c concept is to align irrigation scheduling and amounts with crop requirement and soil characteristics. This is more readily achievable in broadacre cropping systems with largely homogeneous crop canopies, and in which the whole farm needs to be watered during irrigation. It is less appropriate for widely spaced trees or vines in orchards that even at the peak of vegetative development do not fully cover the ground surface. For orchards significant ground surface remains bare and to which applying water amounts to wastage.

5.4.4   SIMPLIFYING IRRIGATION SCHEDULING WITH DRIP SYSTEM FOR TREE CROPS

Aligning water application with crop needs is especially difficult for row crops (vines and trees) where the canopy is distributed in discrete patterns with often bare ground in-between. For the full advantage of drip technology in orchard systems to be realized reliable estimates of ET are critical. For trees, shrubs and vines sapflow have proven reliable in estimating ET [8, 19] or micrometeorological [16] techniques. Both of these techniques present difficulties in converting raw data into meaningful quantities of water-use for scheduling irrigation with drip systems.

Effective and efficient irrigation scheduling that avoids many of the uncertainties is possible by simply approximating optimum crop water-use (ET) with a surrogate obtained by using fraction of canopy cover to scale local ET_o. The fraction of groundcover can be effectively used as putative fraction of solar radiation intercepted by the vegetation. Fractional canopy cover can be easily estimated from tree density and dimensions, and it is a parameter with both physiological and management significance [5].

TABLE 5.2   Crop and Water Use Characteristics for Four Olive Groves During the 2000–2001 Irrigated Growing Season (September–May) in South Australia [16]

Features	Waikerie	Two Wells	Balaklava	Grenock
Grove Characteristics				
Location	34° 9'S;	34° 34'S;	34° 7'S;	34° 27'S;
	140° 0'E	138° 22'E	138° 22'E	139° 0'E
Tree spacing (trees/ha)	8 x 5 m² (156)	7 x 7 m² (204)	10.5 x 5.0 m² (190)	15 x 5 m² (129)
Fraction of tree cover (*i*)	0.52	0.40	0.17	0.41
Irrigation system	Full-cover sprinkler	Drippers	Drippers	Rainfed
Groundcover	Mowed grasses and weeds	Mostly bare	Green manure crop in winter	Unmanaged annual weeds
Water Use Characteristics				
Potential evapotranspiration (ET$_o$, mm)	1492	1370	1370	1223
Rainfall (mm)	118	153	171	206
Irrigation (mm)	486	371	318	—
Total water input (mm)	604	524	489	206
Water required for olive tree transpiration (*i*ET$_o$, mm)	687	415	181	418
Crop factor (K$_{ci}$)	0.46	0.30	0.13	0.34
Water requirement based on K$_c$ 0.8 for the grove (ET$_{Kc}$, mm)[1]	1238	1178	1178	1051
Potential water saving. mm = ET$_{kc}$ − *i*ET	551	763	997	633

[1]K$_c$ based on seasonal average for mature olive groves [7].

When used to scale the ET in Eqn. 8 provides putative upper limits of water required for transpiration through the tree canopies (ET_c):

$$ET_c = iET_o \qquad (9)$$

Thus i is a putative estimate of fraction of incident radiation intercepted by the crop canopy, and ET_c then represents water that has to be supplied to avoid water-stress. The ET_o is readily available and routinely published by meteorological authorities around the world.

5.4.5 APPLICATION OF FRACTIONAL GROUNDCOVER IN IRRIGATION SCHEDULING

The concept of fractional groundcover for irrigation management in row crops was explored using a number of olive groves that differed widely in their tree and management characteristics (Table 5.2). Water balance analysis was undertaken to determine ET for these groves as reported previously by Yunusa [16]. Implementation of Eq. (9) produced a wide range in temporal water rates for the upper limits of transpiration, and reflecting the diverse management strategies adopted by the growers that produced the different canopy covers. In all cases late spring (November) to end of summer (February) represented peak transpirational water requirement (Figure 5.5) when it exceeded 45 kL/ha/d at Waikerie compared with just 10 kL/ha/d at Balaklava. Transpirational water requirement increased exponentially with fractional groundcover:

$$y = 96.118e^{3.6992x}, r^2 = 0.9947 \qquad (10)$$

Thus while the difference in the seasonal transpirational water requirement between Two Wells and Balaklava was about 40%, which was of the similar magnitude as the difference in their fractional canopy cover, the 24% difference in the canopy cover between the groves at Waikerie and Two Wells induced as much as 40% difference in their seasonal transpirational water requirement.

In both dip irrigated groves at Two Wells and Balaklava the amount of water input was sufficient to meet that required for transpiration by the

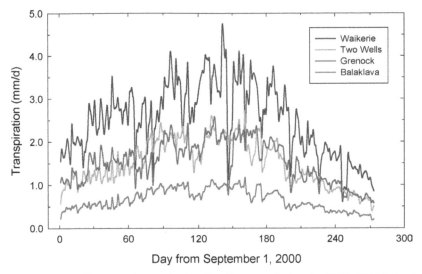

FIGURE 5.5   Daily rates of transpiration for olive trees during the 2000–2001 irrigation season in groves at Waikerie, Two Wells, Grenock and Balaklava in South Australia.

olive trees (Table 5.2). The crop coefficient for satisfying the tree water requirement (K_{ci}) in the two groves was well at the low end of the range (0.15–0.85) commonly reported for irrigated olives [9, 13]. The advantage of scaling water requirement with canopy cover becomes apparent when seasonal water requirements are based on Kc of 0.86 found for mature olives under non-water stressed conditions in Spain [7]. For all the four olive groves, at least 500 mm (0.5 ML/ha) of water could be saved when fractional canopy was taken into account (Table 5.2). This represented more than half of what would have been applied if the conventional K_c was applied.

To take full advantage of drip systems therefore, irrigation water requirements should aim to meet demand by the trees rather than the whole of the orchard or the grove. This means that the commonly used crop factor is somewhat anachronistic when simply applied to row crops with incomplete canopy cover. Drip systems that deliver water to the plant with little or no wetting of the groundcover and/or bare ground needs to be matched with a system that restricts water delivery to other surfaces in the orchard. By restricting water supply to the tree crop, drip systems ensures that transpiration that drives physiological processes, including fruit production, would be fully supported.

5.5   CONCLUSIONS

In this chapter, we have demonstrated that the coefficient of variation for the nPC dripper model evaluated was 2.59, 2.41 and 4.34% for the 2, 4 and 8 dripper (nPC) model respectively. In the case of the PC model, the minimum coefficient of variation was approximately 5% and appeared to increase with pressure from the 150 kPa pressure point. Also, efficient management of dripper systems in orchards requires a capacity to match water application with crop requirements. This can be achieved by scaling evaporative demand by the fraction of canopy cover, since the drip system supplies water directly within the confined space containing the bulk of the tree/vine root system. Advantages of such a system include substantial water saving and improved water-use efficiency. It is concluded that for drip irrigated orchards, irrigation schedule should aim to meet transpirational water demand for trees rather than evapotranspiration for the whole orchard.

5.6   SUMMARY

As compared to the conventional pressured irrigation methods, drip systems are generally considered to be more water and labor efficient, especially for row crops such as trees and vines. This is largely because in this system water drips to a confined space within the vicinity of the plant roots at low flow rates. This chapter focuses on the field and laboratory evaluation of drip systems as an integral part of assessing the efficacy of the technology in terms of optimizing water-use efficiency and minimizing water losses in irrigated cropping systems. The uniformity of flow rate and the flow rate-pressure profiles for a sample of pressure compensating (PC) and non-pressure-compensating (nPC) drippers were evaluated in this study. The capacity of drip system to supply water directly close to a restricted target areas (root zone) makes it particularly suited for orchard irrigation allowing water supply to individual tree/vine system be effectively managed. Using olive groves in South Australia as examples, we demonstrated how water-use efficiency can be optimized through water saving for tree/vine orchards. By simply scaling the commonly used crop coefficient concept with the fraction of canopy cover temporal and seasonal water requirements can be approximated that is consistent with transpirational potential of the crops. Significant water savings can be achieved with drip irrigation system when matched with potential water demand of trees rather than the whole of the orchard.

KEYWORDS

- coefficient of variation
- crop factor
- drip irrigation
- drippers
- efficient
- evaluation
- field
- flow rate
- groundcover
- hydraulics
- irrigation scheduling
- laboratory
- orchards
- pan evaporation
- potential evapotranspiration
- pressure
- reference evapotranspiration
- uncertainty in measurements
- uniformity

REFERENCES

1. Allen, R. G., & Fisher, D. K. (1990). Low-cost electronic weighing lysimeters. *Transactions of the ASAE*, 33, 1823–1833.
2. Allen, R. G., Pereira, L. S., Raes, D., & Smith, M. (1998). *Crop Evapotranspiration-Guidelines for Computing Crop Water Requirements*. FAO Irrigation and drainage paper 56. FAO, Rome.

3. ASAE, (2003). EP405.1: Design and installation of micro irrigation systems. In *ASAE: Standards*. St. Joseph, MI.

4. Bentley, R. (2005). Uncertainty in measurement: the ISO guide. *National Measurement Institute*, Commonwealth of Australia, Canberra.

5. Connor, D. J. (2005). Adaptation of olive (*Olea europaea* L.) to water-limited environments. *Crop and Pasture Science*, 56, 1181–1189.

6. Dutta, D. (2009). *Characterization of drip emitters and computing distribution uniformity in a drip irrigation system at low pressure under uniform land slopes*. Master Thesis, Texas A&M University.

7. Fernández Luque, J. E. (2006). Irrigation Management in Olive. http://www.irriqual.eu/documentos/fern%C3%A1ndez%20olive.pdf.

8. Giorio, P., & Giorion, G. (2003). Sap flow of several olive trees estimated with the heat-pulse technique by continuous monitoring of a single gauge. *Environmental and Experimental Botany*, 49, 9–20.

9. Goldhammer, D. A., Dunai, J., & Fergusson, L. (1993). Water use of Manzanillo olives and responses to sustained deficit irrigation. *Acta Hort.*, 335, 365–371.

10. Goyal, Megh, R. (2015). *Research Advances in Sustainable Micro irrigation, Volumes 1 to 10.* Oakville, ON, Canada: Apple Academic Press Inc.

11. ISO 9261, (2004). Agricultural irrigation equipment – emitters and emitting pipe – specification and test methods. *International Organization of Standardization*, Geneva, Switzerland.

12. Karmeli, D. (1977). Classification of Flow Regime Analysis of Drippers. *Journal of Agricultural Engineering Research*, 22, 165–173.

13. Testi, L., Villalobos, F. J., & Orgaz, F. (2004). Evapotranspiration of young irrigated olive orchard in southern Spain. *Agricultural and Forest Meteorology*, 121, 1–18.

14. WATERWISE, (2005). Evaluating your pressurized system: Systems – Drip (trickle) systems. *NSW Department of Primary Industries*.

15. Wu, I. (1997). An assessment of hydraulic design of micro-irrigation systems. *Agricultural Water Management,* 32(3), 275–284.

16. Yunusa, I. A. M., Zeppel, M. J. B., & Nuberg, I. K. (2008). Water use efficiency reflects management practices in Australian olive groves. *Journal of Horticultural Science and Biotechnology*, 83, 232–238.

17. Yunusa, I. A. M., Sedgley, R. H., & Tennant, D. (1992). Dynamics of water use under annual legume pastures in a semi-arid Mediterranean environment. *Agricultural Water Management*, 22, 291–306.

18. Yunusa, I. A. M., Sedgley, R. H., & Tennant, D. (1994). Evaporation from bare soil in south-western Australia-a test of two models using lysimetry. *Soil Research*, 32, 437–446.

19. Yunusa, I. A. M., Walker, R. R., Loveys, B. R., & Blackmore, D. H. (2000). Determination of transpiration in irrigated grapevines: comparison of the heat-pulse technique with gravimetric and micrometeorological methods. *Irrigation Science*, 20, 1–8.

20. Zhao, H., Xu, D., & Gao, B. (2014). Uncertainty Assessment of Measurement in Variation Coefficient of Drip Irrigation Emitters Flow rate. *IEEE Transactions on Instrumentation and Measurement*, 63(4), 805–812.

CHAPTER 6

FERTIGATION AND GROUND WATER QUALITY WITH TREATED WASTEWATER EFFLUENT

C. PRABAKARAN

CONTENTS

6.1   INTRODUCTION

Intensive cultivation of crops with surface irrigation along with heavy surface application of fertilizers has affected the environment in several ways (including ground water depletion and pollution with nitrates) and created economic problems in different areas of the world. It is estimated by FAO that the annual increase rate of world fertilizer consumption in the period of 2008–2013 is 2.2% for N, 3.8% for P_2O_5 and 5.3% for K_2O.

Forecasts for world-wide fertilizer consumption in 2013 and 2016 are 184 and 194 million tons of nutrients, respectively. Thus, the only reasonable way to solve this problem is to improve fertilizer and water management using advanced management systems like drip fertigation. The demand for irrigation water has gone up due to increased cultivation. The climatic changes and decline in rainfall further added to water woes of the state in the form of inadequate recharging of ground water.

In irrigated agriculture, one of the most practical management methods to apply fertilizers is to inject them directly into the irrigation water. This process is known as fertigation [13]. Some important advantages of fertigation in comparison with traditional fertilizer application methods include flexibility and manageability, cost-effectiveness, the potential for improved fertilizer distribution uniformity and application efficiency (which results in more uniform crop growth along the field), lower losses due to reduced osmotic pressure (low fertilizer concentration), and the possibility to split nutrients application during the growing season. Fertigation can be effectively used to control fertilizer losses and the resulting pollution risk.

The Ballalpur Industrial Packaging Company Limited (BIPCO), located at the foothills of Western Ghats of Thekkampatty Village, Coimbatore District is producing fine quality duplex paper and paperboard from waste papers discharges around 2100–2600 $m^3 d^{-1}$ wastewater which is being used to irrigate about 40 ha of high water requirement crop like banana crop through surface irrigation that led to ground water contamination. Moreover another 60 ha of land is left uncultivated due to lack of water facilities. This area could possibly be brought under cultivation if the crop is drip irrigated with treated effluent without any impact on ground water quality.

This chapter discusses the effects of fertigation on ground water quality under effluent irrigation under drip-irrigated banana.

6.2   LITERATURE REVIEW

Industrialization is believed to cause inevitable problem of pollution of water, soil and air. Pulp and paper industries use large volume of water, the bulk of which is released as effluent requiring proper treatment and disposal. Since these water fall in borderline as saline water, they can be considered as potential source for irrigation [11]. These effluents not only contain nutrients that enhance the growth of crop plants but also have toxic materials especially sodium, which increases the exchangeable sodium percentage (ESP) to harmful level during land disposal. The effect of high ESP is manifested by reduced soil permeability and specific ion toxicity to crops. Further waste water having appreciable concentration of carbonate and bicarbonate alkalinity exhibits a tendency to precipitate calcium in the soil as $CaCO_3$ [18]. Heavy metals tend to accumulate in the soil and plants in undesirable amounts and proportions as a result of disposal of paper mill effluents [37]. Therefore, it is essential that the impact of effluent on crop yield and their effect on soil properties should be assessed before they are recommended for irrigation.

Trickle irrigation has the greatest potential in increasing yields of crops with significant savings in water and nutrients as compared to other conventional methods [26]. Brackish water could effectively be used in drip irrigation. When brackish water is applied frequently under drip irrigation the salinity and sodicity of the soil especially in the root zone of the crop is maintained at low level. Muthuchamy and Valliappan, [27] suggested that the treated paper mill effluent which is saline in nature and devoid of heavy metal pollutants as per State Pollution Control Board (SPCB) norms could possibly be used in drip irrigation for high water requiring crops like banana and sugarcane to alleviate soil salinity hazards in the root zone and to prevent the possible ground water contaminations with organic and inorganic pollutants. Several studies [3, 5, 21, 23, 25, 36] have also indicated that the irrigation water with total salt concentration of 2 g L^{-1} could safely be utilized in drip irrigation.

Fertigation is a technique that combines irrigation with fertilization through any micro irrigation system especially through drip irrigation. Fertigation could bring an accurate control of water and nutrients in the immediate vicinity of the root system. Hence, it is easy and efficient to fertilize the crop and prevents fertilizer contamination of ground water through leaching below the crop root zone [13]. Slow and frequent watering eliminates wide fluctuation of soil moisture under drip irrigation resulting in better growth and yield [3].

Among the fruit crops, banana is well known for its high water requirement, high evaporative demand, high transpiration, shallow root system, poor ability to draw water from soil beneath field capacity and high sensitivity to soil water deficiency. Thus, it requires liberal supplies of irrigation water throughout its life cycle, emphasizing the importance of correct irrigation scheduling. Fertigation has been proved to be of great success in banana in terms of water and labor saving with increased water use efficiency culminating in early cropping and heavy yield [33] and it is an environmentally safer technology which prevents ground water contamination [21].

The advantage of micro-irrigation over surface irrigation is application of water and nutrients to only to the part of the soil volume, where active roots are concentrated, enhances the fertilizer use efficiency and reduces leaching of nutrients to deep ground water by seasonal rains. The main advantage of N micro-fertigation over broadcast N fertilization in orange was reduced nitrate leaching below the soil root volume [3]. The nitrate concentration remained higher in the root zone with frequent trickle irrigation to sweet corn than with flood irrigation. Reduced NO_3 leaching under micro-fertigation with an increase in fruit yield and quality has been recorded in tomato and celery [10].

Hagin and Lowengart [13] reported that intensification of agriculture by irrigation and enhanced use of fertilizers may generate pollution by increased level of nutrients in underground and surface waters. Most of the irrigation is by open system having a relatively low efficiency of water application. A higher efficiency may be gained by pressurized irrigation system. Drip irrigation generates restricted root system requiring frequent nutrient supply that may be satisfied by applying fertilizers in irrigation water (fertigation). Maximization of crop yield and quality and minimization of leaching below the root volume may be achieved by managing fertilizer concentration in measured quantities of irrigation water, according to crop requirement.

Fertigation had invariably increased the efficiency of applied nutrients in banana over manual application [21]. The fertilizer applied in solution form directly to the active root zone in small quantities has been efficiently absorbed and utilized showing reduced possibilities of leaching and utilization by banana plants in a better way than those under conventional system. The efficiency of fertigation has been exhibited by lower nutrient status (N and K) corresponding to higher leaf nutrient status over control during peak vegetative stage and at harvest. On the other hand high volatilization and leaching are commonly associated with conventional system of fertilization wherein large quantity of fertilizer were applied at wider intervals causing every possibilities of leaching below crop root zone leading to ground water contamination of applied nutrients. Hence, conventional fertilizer application in huge quantities at longer intervals can be replaced with fertigation at small quantities at shorter intervals to prevent ground water contamination and maintain soil health. Finally they concluded that fertigation could be taken up as an environmentally safer technology to prevent ground water contamination.

The above foregoing literature reveals that either saline water or treated industrial effluents having salt concentration < 2 g L^{-1} could be used for drip irrigation without any adverse effect on yield and quality of crop produce, soil and ground water. Use of drips for effluent irrigation will increase the area under cultivation, reduce labor consumption, fertilizer losses and prevent soil and ground water contamination. The literatures pertaining to effluent irrigation to banana through drip irrigation are meager. So the present study was proposed.

6.3   MATERIALS AND METHODS

The investigation on the effect of fertigation of treated paperboard mill effluent and solid amendments on ground water quality besides improving soil characteristics, crop growth, quality of crop produce and are carried out at the Bipco Paper Board Industries Pvt. Ltd, Thekkampatti village, (Mettupalayam taluk) and in the Department of Environmental Sciences, TNAU, Coimbatore located in Coimbatore district of Tamil Nadu (India) during May, 2002 to April, 2003.

The experiment was conducted in split plot design with two replications and banana (Robusta) was selected as test crop. The treatments were assigned in main plots and sub plots. Irrigation treatments were assigned in main plot that consisted of seven treatments (I_1 – Farmer's practice as control (Surface Irrigation with river water (RW)+ 100% NK), I_2 – RW + Drip irrigation (DI) + 75% NK thro' fertigation, I_3 – RW + DI + 50% NK thro' fertigation, I_4 – Treated effluent (TE) + Basin irrigation (BI) + 100% NK thro' soil application I_5 – TE + BI + 75% NK thro' soil application, I_6 – TE + DI + 75% NK thro' fertigation, I_7 – TE + DI + 50% NK thro' fertigation). Amendments were applied in Sub-plot. Three types of amendments were applied in the soil (A_1 – Fly ash @ 6 t ha^{-1}, A_2 – Biocompost @ 5 t ha^{-1}, A_3 – Fly ash @ 6 t ha^{-1} + Biocompost @ 5 t ha^{-1} + Green manure @ 6.25 t ha^{-1} recommended 100% NPK is 110: 35:330 g of NPK plant^{-1} y^{-1}). The entire P was applied through single super phosphate as basal dressing in the pit before planting the suckers uniformly for all the treatments.

The experimental area was irrigated with river water obtained from River Bhavani and treated paperboard mill effluent from Bipco according to the treatments. The treated effluent was neutral in reaction with high salinity, contained appreciable amounts of nutrient cations viz., Na, Ca, Mg and anions viz., Cl, SO_4 and HCO_3 with less sodium hazard (SAR <10). The percent sodium was well below the tolerance limit of 60 and the parameters recorded were well within the range of permissible limit prescribed by the Tamil Nadu State Pollution Control Board norms (TNSPCB). The effluent was rich in microbial load with the dominance of bacteria over fungi and actinomycetes. The characteristics of the effluent and river water used for the study were given in the Table 6.1.

The drip system was installed as described by Udayasoorian and Prabakaran [38]. Piezometers were installed as per the procedure described by Latha et al. [20] in each plot to study the ground water quality. Water samples were collected from treated effluent out let and piezometers of the respective plots and the samples for the analysis of dissolved oxygen (DO) were added with one ml of manganese sulfate solution and one ml of alkaline potassium iodide solution. Samples for the determination of Biochemical Oxygen Demand (BOD) were preserved by adding five ml of washed chloroform (Chloroform and distilled water were taken in a separating funnel, shaken well and the water layer was discarded) per liter of the sample [2]. The pH was measured in the spot itself. Carbonates and bicarbonates were analyzed immediately after bringing the sample from field to laboratory. Samples were analyzed for various properties like BOD, COD, total hardness, CO_3, Cl, SO_4, Na, Ca, Mg, K contents and percent sodium values by following standard procedures.

The data on the observation recorded and the characters studied were statistically analyzed by the procedure described by Gomez and Gomez [12] using AGRES software. Wherever the results are significant, the critical difference at 5% level was presented.

TABLE 6.1   Characteristics of Treated Effluent and River Water Used for Irrigation

Characteristics	Treated effluent	River water
pH	7.50	7.05
Electrical conductivity (EC) (dS m^{-1})	1.8	0.05
Biological oxygen demand (BOD) (ppm)	18	4.2
Chemical oxygen demand (COD) (ppm)	90	4.5
Total dissolved solids (TDS), (ppm)	900	47
Total suspended solids (TSS) (ppm)	90	20
NH_4-N (ppm)	155	Nil
NO_3 N (ppm)	42	Nil
Total P (ppm)	1.7	Nil
Total K (ppm)	6.6	1.00
Ca (cmol L^{-1})	11.6	0.84
Mg (cmol L^{-1})	6.03	0.53
Na (cmol L^{-1})	11.5	0.09
SAR	3.87	0.11
CO_3 (ppm)	Nil	Nil
HCO_3 (ppm)	117	1.4
Cl (ppm)	350	37
SO_4 (ppm)	127	42
Bacteria (($\times 10^6$ ml^{-1} *CFU*))	33	7
Fungi ($\times 10^4$ ml^{-1} CFU)	10	5
Actinomycetes ($\times 10^3$ ml^{-1} CFU)	20	2

6.4   RESULTS AND DISCUSSION

The pH values of the ground water samples collected from the piezometer varied from 6.59 to 7.42 at harvest stage (Table 6.2). Neither the irrigation treatments nor the amendments and their interactions were effective in influencing significant changes on pH of the ground water.

Significant increase in ground water EC (Table 6.3) was observed in effluent irrigation treatments over river water irrigation. It varied from 0.02 to 1.75 dS m^{-1} at harvest stage. Among the irrigation treatments, basin irrigation of effluent (I_4, I_5) significantly increased the EC, while river water fertigation (I_2, I_3) on par with I_1 and I_6 significantly reduced it. Among the amendments, lower EC values were recorded in combined application of

TABLE 6.2 Effect of Effluent Irrigation and Amendments on pH of Piezometer Water Samples at Harvest Stage

Irrigation (I)/Amendments (A)	A_1	A_2	A_3	Mean
I_1 – Farmer's practice (surface irrigation)	7.31	7.00	6.90	**7.07**
I_2 – River water fertigation with 75% NK	7.00	7.00	7.00	**7.00**
I_3 – River water fertigation with 50% NK	7.01	7.00	6.99	**7.00**
I_4 – Effluent basin irrigation with 100% NK	7.42	7.33	7.31	**7.35**
I_5 – Effluent basin irrigation with 75% NK	7.33	7.30	6.59	**7.07**
I_6 – Effluent fertigation with 75% NK	7.01	7.00	7.12	**7.04**
I_7 – Effluent fertigation with 50% NK	7.12	7.02	7.00	**7.05**
Mean	**7.17**	**6.99**	**7.09**	
	I	**A**	**IxA**	**A X I**
SEd	0.26	0.19	0.48	0.49
CD (0.05)	NS	NS	NS	NS

(A_1 – Fly ash @ 6 t ha^{-1}; A_2 – Biocompost @ 5 t ha^{-1} and A_3 – Fly ash @ 6 t ha^{-1} + Biocompost @ 5t ha^{-1} + green manure @ 6.25 t ha^{-1}).

TABLE 6.3 Effect of Effluent Irrigation and Amendments on EC (dS m^{-1}) of Piezometer Water Samples at Harvest Stage

Irrigation (I)/Amendments (A)	A_1	A_2	A_3	Mean
I_1 – Farmer's practice (Surface irrigation)	0.13	0.11	0.06	**0.10**
I_2 – River water fertigation with 75% NK	0.02	0.02	0.02	**0.02**
I_3 – River water fertigation with 50% NK	0.02	0.02	0.02	**0.02**
I_4 – Effluent basin irrigation with 100% NK	1.75	1.40	0.84	**1.33**
I_5 – Effluent basin irrigation with 75% NK	1.64	1.31	0.78	**1.24**
I_6 – Effluent fertigation with 75% NK	0.21	0.17	0.11	**0.16**
I_7 – Effluent fertigation with 50% NK	0.32	0.25	0.15	**0.24**
Mean	**0.58**	**0.47**	**0.29**	
	I	**A**	**I at A**	**A at I**
SEd	0.07	0.04	0.10	0.09
CD (0.05)	0.17	0.08	0.23	0.20

(A_1 – Fly ash @ 6 t ha^{-1}; A_2 – Biocompost @ 5 t ha^{-1} and A_3 – Fly ash @ 6 t ha^{-1} + Biocompost @ 5 t ha^{-1} + green manure @ 6.25 t ha^{-1})

fly ash + compost + green manure (A_3), while higher values were recorded in fly ash alone (A_1) The interaction effects were significant.

The BOD of piezometer water samples significantly increased due to basin irrigation of effluent, while it was decreased due to fertigation treatments either through effluent or river water (Table 6.4). Application of fly ash (A_1) alone increased the BOD values, whereas it was decreased due to incorporation of fly ash + compost + green manure (A_3).

Fertigation treatments decreased the COD of ground water samples. It ranged from 52 to 443 ppm at harvest stages (Table 6.5). Among the irrigation treatments, fertigation either through river water (I_2, I_3) or effluent (I_6, I_7) significantly reduced the COD values, while basin irrigation of effluent (I_4, I_5) increased the COD values and it was on par with farmer's practice (I_1). Addition of fly ash (A_1) alone increased the COD compared to rest of the amendments. Interaction between irrigation treatments and amendments was not significant.

The Ca content of ground water sample at harvest stage varied from 0.04 to 3.85 cmol L^{-1} (Table 6.6). Effluent irrigation, amendments or interaction had non-significantly influenced the Ca content.

The Mg content of ground water samples varied from 0.02 to 4.12 cmol L^{-1} (Table 6.7). Similar to Ca content of piezometer water

TABLE 6.4 Effect of Effluent Irrigation and Amendments on BOD (ppm) of Piezometer Water Samples at Harvest Stage

Irrigation (I)/Amendments (A)	A_1	A_2	A_3	Mean
I_1 – Farmer's practice (Surface irrigation)	48.4	39.9	35.7	**41.3**
I_2 – River water fertigation with 75% NK	8.4	8.4	6.3	**7.7**
I_3 – River water fertigation with 50% NK	10.5	10.5	8.4	**9.8**
I_4 – Effluent basin irrigation with 100% NK	54.1	45.1	40.5	**46.6**
I_5 – Effluent basin irrigation with 75% NK	50.3	42.1	37.5	**43.3**
I_6 – Effluent fertigation with 75% NK	10.5	10.5	8.4	**9.8**
I_7 – Effluent fertigation with 50% NK	8.4	8.4	6.3	**7.7**
Mean	**27.2**	**23.6**	**20.4**	
	I	A	I at A	A at I
SEd	2.7	1.2	3.7	3.1
CD (0.05)	6.7	2.5	NS	NS

(A_1 – Fly ash @ 6 t ha^{-1}; A_2 – Biocompost @ 5 t ha^{-1} and A_3 – Fly ash @ 6 t ha^{-1} + Biocompost @ 5 At ha^{-1} + green manure @ 6.25 t ha^{-1})

TABLE 6.5 Effect of Effluent Irrigation and Amendments on COD (ppm) of Piezometer Water Samples at Harvest Stage

Irrigation (I)/Amendments (A)	A_1	A_2	A_3	Mean
I_1 – Farmer's practice (Surface irrigation)	397	328	293	**339**
I_2 – River water fertigation with 75% NK	69	69	52	**63**
I_3 – River water fertigation with 50% NK	86	86	69	**80**
I_4 – Effluent basin irrigation with 100% NK	443	369	332	**382**
I_5 – Effluent basin irrigation with 75% NK	413	345	308	**355**
I_6 – Effluent fertigation with 75% NK	86	86	69	**80**
I_7 – Effluent fertigation with 50% NK	69	69	52	**63**
Mean	**223**	**193**	**168**	
	I	A	I at A	A at I
SEd	23.0	9.6	31.0	25.0
CD (0.05)	55.0	21.0	NS	NS

(A_1 – Fly ash @ 6 t ha^{-1}; A_2 – Biocompost @ 5 t ha^{-1} and A_3 – Fly ash @ 6 t ha^{-1} + Biocompost @ 5 t ha^{-1} + green manure @ 6.25 t ha^{-1})

TABLE 6.6 Effect of Effluent Irrigation and Amendments on Ca (cmol L^{-1}) of Piezometer Water Samples at Harvest Stage

Irrigation (I)/Amendments (A)	A_1	A_2	A_3	Mean
I_1 – Farmer's practice (Surface irrigation)	0.29	0.24	0.13	**0.22**
I_2 – River water fertigation with 75% NK	0.04	0.04	0.04	**0.04**
I_3 – River water fertigation with 50% NK	0.04	0.04	0.04	**0.04**
I_4 – Effluent basin irrigation with 100% NK	3.85	3.08	1.85	**2.93**
I_5 – Effluent basin irrigation with 75% NK	3.61	2.88	1.72	**2.74**
I_6 – Effluent fertigation with 75% NK	0.46	0.37	0.24	**0.36**
I_7 – Effluent fertigation with 50% NK	0.70	0.55	0.33	**0.53**
Mean	**1.28**	**1.03**	**0.62**	
	I	A	I at A	A at I
SEd	1.45	0.85	2.30	2.20
CD (0.05)	NS	NS	NS	NS

(A_1 – Fly ash @ 6 t ha^{-1}; A_2 – Biocompost @ 5 t ha^{-1} and A_3 – Fly ash @ 6 t ha^{-1} + Biocompost @ 5 t ha^{-1} + green manure @ 6.25 t ha^{-1})

samples Mg content was non-significantly different due to irrigation treatments, amendments or its interaction.

TABLE 6.7 Effect of Effluent Irrigation and Amendments on Mg (cmol L^{-1}) of Piezometer Water Samples at Harvest Stage

Irrigation (I)/Amendments (A)	A_1	A_2	A_3	Mean
I_1 – Farmer's practice (Surface irrigation)	0.13	0.11	0.06	**0.10**
I_2 – River water fertigation with 75% NK	0.02	0.02	0.02	**0.02**
I_3 – River water fertigation with 50% NK	0.02	0.02	0.02	**0.02**
I_4 – Effluent basin irrigation with 100% NK	4.12	3.30	1.98	**3.13**
I_5 – Effluent basin irrigation with 75% NK	3.85	3.07	1.85	**2.92**
I_6 – Effluent fertigation with 75% NK	0.21	0.17	0.11	**0.16**
I_7 – Effluent fertigation with 50% NK	0.34	0.27	0.17	**0.26**
Mean	**1.24**	**0.99**	**0.60**	
	I	A	I at A	A at I
SEd	1.50	0.72	2.23	2.18
CD (0.05)	NS	NS	NS	NS

(A_1 – Fly ash @ 6 t ha^{-1}; A_2 – Biocompost @ 5 t ha^{-1} and A_3 – Fly ash @ 6 t ha^{-1} + Biocompost @ 5 t ha^{-1} + green manure @ 6.25 t ha^{-1})

Effluent irrigation significantly increased the ground water Na content than river water irrigation (Table 6.8). Among the effluent irrigation treatments, basin irrigation of the effluent (I_4, I_5) recorded higher values of Na than effluent fertigation. Application of fly ash (A_1) increased the Na content in ground water and combined application of fly ash + biocompost + green manure reduced the Na content. The interaction effect was also significant.

The K content of ground water samples at harvest stage ranged from 0.01 to 2.75 cmol L^{-1} (Table 6.9). The same trend as above in Na was observed here also in irrigation treatments and amendments and interaction.

Basin irrigation with effluent recorded higher SSP values than fertigation treatments either through effluent or river water indicating the possibilities of polluting ground water. The SSP of water samples collected in the piezometer varied from 43.54 to 80.31 (Table 6.10). The SSP of ground water samples was not significantly influenced by the incorporation of amendments and their interactions with irrigation treatments.

The ground water chloride content was significantly increased due to effluent irrigation compared to river water irrigation (Table 6.11). Among the effluent treatments, fertigation with effluent reduced the Cl contamination than basin irrigation. Among the amendments, incorporation of

TABLE 6.8   Effect of Effluent Irrigation and Amendments on Na (cmol L^{-1}) of Piezometer Water Samples at Harvest Stage

Irrigation (I)/ Amendments (A)	A$_1$	A$_2$	A$_3$	Mean
I$_1$ – Farmer's practice (Surface irrigation)	0.19	0.17	0.11	**0.16**
I$_2$ – River water fertigation with 75% NK	0.04	0.02	0.02	**0.03**
I$_3$ – River water fertigation with 50% NK	0.04	0.04	0.02	**0.04**
I$_4$ – Effluent basin irrigation with 100% NK	6.18	4.94	2.96	**4.70**
I$_5$ – Effluent basin irrigation with 75% NK	5.76	4.63	2.78	**4.69**
I$_6$ – Effluent fertigation with 75% NK	0.34	0.27	0.17	**0.26**
I$_7$ – Effluent fertigation with 50% NK	0.50	0.40	0.23	**0.38**
Mean	**1.87**	**1.50**	**0.90**	
	I	A	I at A	A at I
SEd	0.23	0.12	0.35	0.33
CD (0.05)	0.57	0.27	0.81	0.70

(A$_1$ – Fly ash @ 6 t ha^{-1}; A$_2$ – Biocompost @ 5 t ha^{-1} and A$_3$ – Fly ash @ 6 t ha^{-1} + Biocompost @ 5 t ha^{-1} + green manure @ 6.25 t ha^{-1})

TABLE 6.9   Effect of Effluent Irrigation and Amendments on K (cmol L^{-1}) of Piezometer Water Samples at Harvest Stage

Irrigation (I)/ Amendments (A)	A$_1$	A$_2$	A$_3$	Mean
I$_1$ – Farmer's practice (Surface irrigation)	0.11	0.08	0.06	**0.08**
I$_2$ – River water fertigation with 75% NK	0.02	0.02	0.01	**0.02**
I$_3$ – River water fertigation with 50% NK	0.02	0.02	0.02	**0.02**
I$_4$ – Effluent basin irrigation with 100% NK	2.75	0.34	1.66	**1.58**
I$_5$ – Effluent basin irrigation with 75% NK	2.59	0.32	1.56	**1.49**
I$_6$ – Effluent fertigation with 75% NK	0.19	0.15	0.08	**0.14**
I$_7$ – Effluent fertigation with 50% NK	0.27	0.23	0.13	**0.21**
Mean	**0.85**	**0.17**	**0.50**	
	I	A	I at A	A at I
SEd	0.08	0.05	0.14	0.14
CD (0.05)	0.21	0.11	0.31	0.29

(A$_1$ – Fly ash @ 6 t ha^{-1}; A$_2$ – Biocompost @ 5 t ha^{-1} and A$_3$ – Fly ash @ 6 t ha^{-1} + Biocompost @ 5 t ha^{-1} + green manure @ 6.25 t ha^{-1})

fly ash + compost + green manure (A$_3$) decreased the chloride content, whereas addition of fly ash increased the chloride content of ground water

TABLE 6.10   Effect of Effluent Irrigation and Amendments on SSP Values of Piezometer Water Samples at Harvest Stage

Irrigation (I)/ Amendments (A)	A_1	A_2	A_3	Mean
I_1 – Farmer's practice (Surface irrigation)	51.63	52.86	56.24	**53.58**
I_2 – River water fertigation with 75% NK	60.66	43.54	43.54	**49.24**
I_3 – River water fertigation with 50% NK	60.66	60.66	43.54	**54.95**
I_4 – Effluent basin irrigation with 100% NK	80.19	80.18	80.18	**80.18**
I_5 – Effluent basin irrigation with 75% NK	80.10	80.18	80.31	**80.20**
I_6 – Effluent fertigation with 75% NK	54.02	54.11	52.86	**53.67**
I_7 – Effluent fertigation with 50% NK	53.63	53.96	53.07	**53.55**
Mean	**49.23**	**49.24**	**49.09**	
	I	A	I at A	A at I
SEd	1.1	21.0	22.1	22.1
CD (0.05)	2.2	NS	NS	NS

(A_1 – Fly ash @ 6 t ha^{-1}; A_2 – Biocompost @ 5 t ha^{-1} and A_3 – Fly ash @ 6 t ha^{-1} + Biocompost @ 5 t ha^{-1} + green manure @ 6.25 t ha^{-1})

samples. The interaction effect was found to be significant. The treatment combination I_4A_1 recorded the highest value of 9.48 cmol L^{-1} of chloride and it was on par with I_5A_1. The least values were recorded under I_2A_3.

The sulfate content of ground water samples varied from 0.10 to 3.74 cmol L^{-1} (Table 6.12). The same trend as that of Cl was observed here also where as the magnitude of SO$_4$ contamination was being less compared to Cl.

Fertigation treatments lowered ground water nitrate pollution than surface irrigation. The magnitude of NO$_3$ pollution was higher in effluent basin irrigation (I_4, I_5) than rest of the irrigation treatments. Addition of fly ash + compost + green manure (A_3) reduced the ground water nitrate pollution and fly ash alone increased the NO$_3$ pollution and the values ranged from 7.6 to 79.2 ppm (Table 6.13).

In general, fertigation treatments (I_6, I_7, I_2, I_3) recorded higher yield than basin irrigation treatments (I_4, I_5) and farmer's practice (I_1). The yield obtained from the field trial varied from 21.5 to 57.5 kg plant^{-1} (Table 6.14). The yield was increased (57.5 kg plant^{-1}) due to effluent fertigation with 75% NK (I_6) was on par with other fertigation treatments. Application of amendments or interaction did not show any differences in yield.

TABLE 6.11   Effect of Effluent Irrigation and Amendments on Cl (cmol L^{-1}) of Piezometer Water Samples at Harvest Stage

Irrigation (I)/Amendments (A)	A$_1$	A$_2$	A$_3$	Mean
I$_1$ – Farmer's practice (Surface irrigation)	2.96	2.37	1.78	**2.37**
I$_2$ – River water fertigation with 75% NK	0.59	0.59	0.00	**0.39**
I$_3$ – River water fertigation with 50% NK	0.59	0.59	0.59	**0.59**
I$_4$ – Effluent basin irrigation with 100% NK	9.48	7.76	4.68	**7.30**
I$_5$ – Effluent basin irrigation with 75% NK	8.88	7.28	4.38	**6.85**
I$_6$ – Effluent fertigation with 75% NK	5.33	4.15	2.37	**3.95**
I$_7$ – Effluent fertigation with 50% NK	5.13	4.34	2.37	**3.95**
Mean	**4.71**	**3.87**	**2.31**	
	I	A	I at A	A at I
SEd	16.9	6.9	22.7	18.4
CD (0.05)	41.4	14.9	52.5	39.6

(A$_1$ – Fly ash @ 6 t ha^{-1}; A$_2$ – Biocompost @ 5 t ha^{-1} and A$_3$ – Fly ash @ 6 t ha^{-1} + Biocompost @ 5 t ha^{-1} + green manure @ 6.25 t ha^{-1})

TABLE 6.12   Effect of Effluent Irrigation and Amendments on SO$_4$ (cmol L^{-1}) of Piezometer Water Samples at Harvest Stage

Irrigation (I)/Amendments (A)	A$_1$	A$_2$	A$_3$	Mean
I$_1$ – Farmer's practice (Surface irrigation)	1.17	0.93	0.70	**0.93**
I$_2$ – River water fertigation with 75% NK	0.23	0.23	0.10	**0.19**
I$_3$ – River water fertigation with 50% NK	0.23	0.23	0.23	**0.23**
I$_4$ – Effluent basin irrigation with 100% NK	3.74	3.06	1.85	**2.88**
I$_5$ – Effluent basin irrigation with 75% NK	3.50	2.87	1.73	**2.70**
I$_6$ – Effluent fertigation with 75% NK	2.10	1.64	0.93	**1.56**
I$_7$ – Effluent fertigation with 50% NK	2.02	1.71	0.93	**1.55**
Mean	**1.86**	**1.52**	**0.92**	
	I	A	I at A	A at I
SEd	0.19	0.08	0.25	0.21
CD (0.05)	0.46	0.17	0.58	0.44

(A$_1$ – Fly ash @ 6 t ha^{-1}; A$_2$ – Biocompost @ 5 t ha^{-1} and A$_3$ – Fly ash @ 6 t ha^{-1} + Biocompost @ 5 t ha^{-1} + green manure @ 6.25 t ha^{-1})

TABLE 6.13   Effect of Effluent Irrigation and Amendments on NO_3 (ppm) of Piezometer Water Samples at Harvest Stage

Irrigation (I)/Amendments (A)	A_1	A_2	A_3	Mean
I_1 – Farmer's practice (Surface irrigation)	42.1	33.6	25.2	**33.6**
I_2 – River water fertigation with 75% NK	8.4	8.4	7.6	**8.1**
I_3 – River water fertigation with 50% NK	8.4	8.4	8.4	**8.4**
I_4 – Effluent basin irrigation with 100% NK	79.2	64.8	39.1	**61.0**
I_5 – Effluent basin irrigation with 75% NK	74.2	60.8	36.6	**57.2**
I_6 – Effluent fertigation with 75% NK	38.0	29.6	16.9	**28.2**
I_7 – Effluent fertigation with 50% NK	36.6	31.0	16.9	**28.2**
Mean	**41.0**	**33.8**	**21.5**	
	I	A	I at A	A at I
SEd	3.8	1.6	5.2	4.4
CD (0.05)	9.4	3.5	12	9.3

(A_1 – Fly ash @ 6 t ha⁻¹; A_2 – Biocompost @ 5 t ha⁻¹ and A_3 – Fly ash @ 6 t ha⁻¹ + Biocompost @ 5 t ha⁻¹ + green manure @ 6.25 t ha⁻¹)

TABLE 6.14   Effect of Effluent Irrigation and Amendments on Fruit Yield (kg Plant⁻¹)

Irrigation (I)/ Amendments (A)	A_1	A_2	A_3	Mean
I_1 – Farmer's practice (Surface irrigation)	34.0	30.5	38.0	**34.2**
I_2 – River water fertigation with 75% NK	53.0	51.5	52.5	**52.3**
I_3 – River water fertigation with 50% NK	51.5	50.5	53.5	**51.8**
I_4 – Effluent basin irrigation with 100% NK	25.5	22.5	28.0	**25.3**
I_5 – Effluent basin irrigation with 75% NK	24.0	21.5	24.0	**23.2**
I_6 – Effluent fertigation with 75% NK	55.0	56.0	57.5	**56.2**
I_7 – Effluent fertigation with 50% NK	56.5	54.5	56.5	**55.8**
Mean	**42.8**	**41.0**	**44.3**	42.7
	I	A	I at A	A at I
SEd	5.7	3.0	8.6	5.7
CD (0.05)	14.1	NS	NS	14.1

(A_1 – Fly ash @ 6 t ha⁻¹; A_2 – Biocompost @ 5 t ha⁻¹ and A_3 – Fly ash @ 6 t ha⁻¹ + Biocompost @ 5 t ha⁻¹ + green manure @ 6.25 t ha⁻¹)

6.4.1 DISCUSSIONS

The result obtained due to analysis of piezometer water samples collected during harvest stage are discussed here under. The pH is the negative logarithm of the hydrogen ion concentration or simply the log of the reciprocal of the hydrogen ion concentration and indicates the degree of acidity or alkalinity of water. In the present investigation, three was no drastic change in pH of water collected in the piezometer. Due to irrigation sources amendments or interaction, this might be due to buffering capacity of soil that might have prevented drastic change of ground water pH.

The concentration of soluble salts in water can be measured in terms of electric conductivity. In the present investigation surface irrigation of treated effluent with 100% NK increased the EC of ground water (Figure 6.1). The reason was due to surface application of higher amount of effluent that might have reached the ground water through percolation and seepage. Among the amendments, application of fly ash increased the EC. The EC was decreased in the plots applied with fly ash + biocompost + green manure. It was due to addition of organic matter by green manure

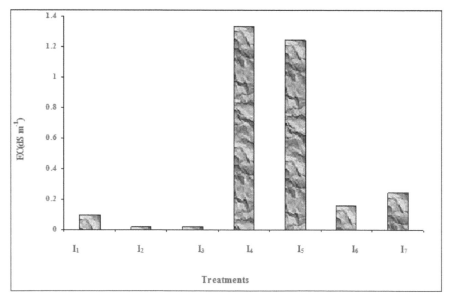

FIGURE 6.1   Effect of irrigation sources on electrical conductivity (EC) of ground water at harvest.

that prevented the downward movement of salts. Increase in ground water EC might be due to increase in concentration of salts like Ca, Mg, Na, K, etc., present in the seepage.

The BOD is defined as the amount of oxygen required by bacteria for stabilizing decomposable organic matter under aerobic conditions. In the present investigation surface irrigation of effluent increased the ground water BOD. It might be due to eutrophication of the ground water with nutrients present in the effluent and applied nutrients. Similarly application fly ash alone increased the BOD, while decreased BOD was recorded in the plots applied with fly ash + biocompost + green manure. This might be due to slow release of applied nutrients and continuous uptake by crop might have reduced the nutrient content of ground water collected in the piezometers.

The chemical oxygen demand (COD) is a measure of oxygen equivalent to that portion of organic matter present in the wastewater sample that is susceptible to oxidation by potassium dichromate. This is an important and quickly measured parameter for stream, sewage and industrial waste samples to determine their pollution strength. Among irrigation treatments, higher COD values were recorded in surface irrigation treatments. Similarly, application of fly ash alone increased the COD.

The cations viz., Ca, Mg, Na and K were increased due to surface irrigation of effluent with application of fly ash alone, while river water irrigation along with combined application of fly ash + biocompost + green manure decreased cations of ground water. Increased cations and anions due to surface irrigation of effluent was reported by Elayarajan [9]. Decrease in ions due to combined application of amendments might be due to precipitation of Ca and Mg by green manure.

Among the irrigation treatments, surface irrigation of effluent with 100% NK increased the Na (Figure 6.2) and K content. Increase in Na and K content might be due to their high mobility that favored to contaminate the ground water easily when compared to other constituents. Similarly, application of fly ash alone increased the Na and K content. It might be due to high Ca and Mg content of the fly ash which had replaced the native Na leach out to ground water.

Among the irrigation sources, surface irrigation of the treated effluent increased the chloride content. Among the amendments, application of fly ash alone increased the chloride concentration. This was due to higher

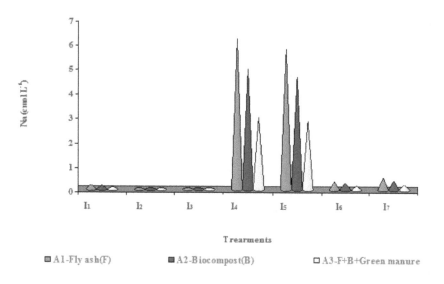

FIGURE 6.2   Effect of irrigation sources on Na content of ground water at harvest.

chloride content of the treated effluent. Increase in chloride content in the ground water samples collected in and around the continuously effluent irrigated area was reported by Elayarajan [9].

Very frequently ground water contain high amount of nitrate. When such type of irrigation water is applied on soils continually, various physical properties will be affected badly, which causes poor growth of plants. In the present investigation, among the irrigation sources, surface application of treated effluent increased the concentration of nitrate (Figure 6.3). Similarly, application of fly ash increased the concentration. The permissible limit for safe drinking water is 50 ppm above which this may cause "blue baby disease." In the study, surface application of treated effluent with fly ash and surface application of treated effluent with bio-compost at harvest stage exceeded the critical limit of 50 ppm. This indi-cates that surface irrigation of the effluent either with fly ash or biocompost alone should be avoided to protect the wells from nitrate pollution. The result indicates that fertigation treatments were effective in protecting the ground water from leaching and percolation of effluent water irrigation.

Yield of banana ha^{-1} was increased due to effluent fertigation with 75% NK. Increased yield in the present study might be due to increase in

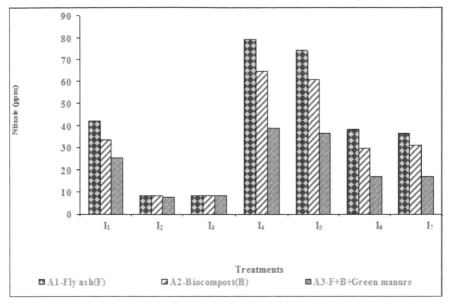

FIGURE 6.3   Effect of irrigation sources on nitrate content of ground water at harvest.

bunch weight plant[-1] [28]. Banana requires very moist conditions for optimum growth and production. A decrease in soil matrix potential adversely affects its performance [14]. Robinson and Alberts [31] and Robinson and Bower [32] have noticed initiation of stress in banana plants at a soil matric potential of around –20 to –25 kpa. As the drip irrigation system maintains very high soil matric potential, at least a part of the root zone without much stress, there were probably ideal conditions exists for better growth and yield of banana compared with basin irrigation where plants are liable to experience increasing stress each day following an irrigation. Increased yield due to fertigation was reported in different crops *viz.*, tomato [4], cotton [7], sweet corn [39], avocado [24], mandarin [35], apricot [30], sugarcane [34], orange [1, 6, 8, 16], papaya [17], hybrid tomato [29], banana [15, 31, 33], and potato [19].

6.5   CONCLUSIONS

The ground water quality parameters (EC, BOD, COD, TDS, Ca, Mg, Na, K, Cl and SO_4) were increased due to basin irrigation of treated board mill effluent. Basin irrigation with effluent recorded higher SSP values and higher nitrate content than fertigation treatments either through effluent or river water indicating the possibilities of polluting ground water source. Surface application of treated effluent with fly ash and surface application of treated effluent with biocompost at harvest stage exceeded the critical limit of 50 ppm.

This indicates that surface irrigation of the effluent either with fly ash or biocompost should be avoided to protect the wells from nitrate pollution in and around Bipco factory area. It is evident that the fertigation treatments, which are more effective in protecting the ground water and reduce leaching and percolation of effluent water in this study.

6.6 SUMMARY

The investigation on the effects of fertigation using treated paperboard mill effluent and soil basal application of solid amendments on ground water quality (besides improving crop growth, yield and quality of banana) under banana cultivation were carried out at the Bipco Paper Board Industries Pvt. Ltd, Thekkampatti village, (Mettupalayam taluk) and in the Department of Environmental Sciences, TNAU, Coimbatore located in Coimbatore district of Tamil Nadu (India) during May, 2002 to April, 2003.

The experiment was conducted in split plot design with two replications. Irrigation treatments were assigned in main plot while amendments were applied as basal in sub plots. Both well water and treated effluent were compared by using drip irrigation system and surface irrigation along with half and three fourth of the recommended dose of NK fertilizers. Main plot treatments were compared with farmers practice (surface irrigation of river water with 100% NK, plant^{-1}y^{-1}) while recommended dose was 110:35:330 g of NPK plant^{-1} y^{-1}. Entire dose of P was applied as basal dressing in the pit before planting the suckers uniformly for all the treatments. Either fly ash (6 t ha^{-1}), biocompost (5 t ha^{-1}) or combination of both with green manure (6.25 t ha^{-1}) were applied as basal in subplots.

Piezometers were installed at center of each plot. Water samples were collected from the piezometers installed in the plot of respective treatments indicated that the ground water quality parameters *viz.,* BOD, COD, EC, TDS, Na, K, Cl, SO_4, NO_3 contents and SSP values were low in drip fertigation treatments while the above contents were higher due to basin irrigation of treated board mill effluent. Moreover, the result indicated basin irrigation of treated effluent with basal application of fly ash or compost alone at harvest stage exceeded the critical limit of 50 ppm. Hence it is recommended to adopt drip fertigation for recycling of treated board mill effluent for banana along with combined basal soil application of fly ash, biocompost and green manure at recommended rates to protect the wells from contamination in and around Bipco factory.

KEYWORDS

- biocompost
- BOD
- Cl
- COD
- drip irrigation
- EC
- effluent
- fertigation
- fly ash
- green manure
- ground water
- ground water pollution
- K
- Na
- NO_3
- SO_4
- Tamil Nadu
- TDS
- treated paper board mill effluent
- water quality

REFERENCES

1. Ananthakumar, A. P., & Bojappa, K. M. (1994). Studies on the effect of drip irrigation on yield and quality of fruits in sweet oranges and economy in water use. Mysore, *J. Agric Sci.*, 28: 338–344

2. Anon, (1965). Standard methods for the examination of water and waste water. *Amer. Water Works Assoc. and Amer. Water Pollution Contd. Federation.* Brosluky, New York, p.600.

3. Baryosef, B. (1999). Advances in fertigation. *Adv. Agron.*, 65: 1–77.

4. Baryosef, B., & Sagiv, B. (1982). Response of tomatoes to N and water applied through trickle irrigation system. I. Nitrogen. *Agron. J.*, 74: 633–636.

5. Bernstein, L., & Francois, L. E. (1975). Effect of frequency of sprinkling with saline waters compared with daily drip irrigation. *Agron. J.*, 67, 185–190.

6. Bharambe, P. R., Mungal, M. S., Shelke, D. K., Oza, S. R., Vaishnava, V. G., & Sondge, V. D. (2001). Effect of soil moisture regimes with drip on spatial distribution of moisture, salts, nutrient availability and water use efficiency of banana. *J. Indian Soc. Soil. Sci.*, 49 (4), 658–665.

7. Bielorai, H., Vaisman, I., & Feign, A. (1984). Drip irrigation of cotton with treated municipal effluents. I. Yield response. *J. Environ. Qual.*, 13(2), 231–238.

8. Dasberg, S., Barakiva, A., Spazisky, S., Cohen, A. (1998). Fertigation versus broad casting in an orange grove. Fertil. Res., 15: 147–154.

9. Elayarajan. M. (2002). Land application of treated paper board mill effluent on soil – water – plant ecosystem. PhD Thesis, Tamil Nadu Agric. Univ., Coimbatore.

10. Feign, A., Letey, J., & Jarrell, W. M. (1982). Celery response to type, amount and method of N fertilizer application under drip irrigation. *Agron. J.*, 74, 971–977.

11. Gomathi, V., & Oblisami, G. (1992). Effect of pulp and paper mill effluent on germination of tree crops. Indian, *J. Environ. Hlth.*, 34: 326–328.

12. Gomez, A. (1984). *Statistical Procedures for Agricultural Research.* II Ed. John Willey and Sons, New Delhi, p. 600.

13. Hagin, J., & Lowngart, G. (1996). Fertigation for minimizing environmental pollution by fertilizers. *Fertl. Res.*, 43, 5–7.

14. Hegde, D. M. (1988). Growth and yield analysis of Robusta banana in relation to soil water potential and N fertilization. Scientia Hortic., 37: 145–155.

15. Hegde, D. M., & Srinivas, K. (1990). Growth, productivity and water use of banana under drip and basin irrigation in relation to evaporation replenishment. Indian, *J. Agron.*, 35 (1&2), 106–112.

16. Idate, G. M., Chandhari, S. M., & More, T. A. (2001). Fertigation in pomegranate. *South Indian Hort.*, *49*: 69–72.

17. Jeyakumar, P., Kumar, N., & Soorianathasundram, K. (2001). Fertigation in papaya. *South Indian Hort.*, 49: 71–75

18. Kannan, K., & Oblisami, G. (1990). Effect of pulp and paper mill effluent irrigation on carbon dioxide evolution in soils. *J. Agron. Crop Sci.*, 164: 116–119.

19. Keshavaiah, K. V., & Kumarasamy, A. S. (1993). Fertigation and water use efficiency in potato under furrow and drip irrigation. *J. Indian Potato Assoc.* 20 (3–4), 210–244.

20. Latha, P., Thangavel, P., & Arulmozhiselvan, K. (2013). Monitoring distillery effluent effect on water quality using piezometer. *Madras Agric, J.* 100(1–3), 98–101.

21. Mahalakshmi, M., Kumar, N., & Soorianatha Sundaram, K. (2002). Fertigation prevents ground water contamination and improves fertilizer use efficiency In: Proc. Nat Sem. *On Strategies on Environment Management,* FC&RI, Mettupalayam, pp. 3–4.

22. Mahalakshmi, M. (2000). *Water and Fertigation Management Studies in Banana.* PhD Thesis, Tamil Nadu Agric. Univ., Coimbatore.

23. Meiri, A., & Plaut, Z. (1985). Crop production and management under saline conditions. *Plant and Soil,* 89: 253–271

24. Michelakis, N., Vougioucalou, E., & Dapaki, G. (1993). Water use, wetted soil volume, root distribution and yield of avocado under drip irrigation. *Agri. Water Manage.,* 24, 119–131.

25. Mizrahi, Y., Taleisnik, E., Kagan-Zur, V., Zohar, Y., Offenbach, R., Matan, E., & Golan, R. (1988). A saline irrigation regime for improving tomato fruit quality without reducing yield. *J. Am. Soc. Hort. Sci.,* 113, 202–205

26. Mohamedharoon, P. (1991). *Effect of Trickle Irrigation and Fertigation on Soil Properties and Nutrient Uptake by Tomato.* PhD Thesis, Tamil Nadu Agric. Univ., Coimbatore.

27. Muthuchamy, I., & Valliappan, K. (1993). Salt dynamics in root zone. *Madras Agric. J.,* 79, 51–52.

28. Prabakaran, C. (2004). *Effect of Treated Board Mill Effluent Fertigation on Yield and Quality of Banana.* PhD Thesis, Submitted to Tamil Nadu Agricultural University. Coimbatore.

29. Prabhakar, M., Vijaya, S., & Naik, C. L. (2001). Fertigation studies in hybrid tomato. *South Indian Hort.,* 49: 98–100.

30. Ranbirsingh, A., Bhandari, R., & Thakur, B. C. (2002). Effect of drip irrigation regimes and plastic mulch on fruit growth and yield of apricot. Indian. *J. Agric. Sci.,* 72 (6), 355–357.

31. Robinson, J. C., & Alberts, A. J. (1986). Growth and yield response of banana to drip irrigation under drought and normal rainfall conditions in the subtropics. *Scientia Hortic.,* 30, 187–202.

32. Robinson, T. C., & Bower, J. P. (1987). A transpiration characteristic of banana leaves in response to progressive depletion of available soil moisture. *Scientia Hortic.,* 31, 289–300.

33. Santhanabosu, S., Rajakrishnamurthy, V., Duraisamy, V. K., & Rajagopal, A. (1995). Studies on the strategy of drip irrigation to banana. *Madras Agric. J.,* 82(1), 44–45.

34. Shinde, S. H., Dahiwalkar, S. D., & Berad, S. M. (2001). Influence of planting technique and fertigation on sugarcane economics and quality. *Indian Sugar,* 4, 17–21.

35. Shirgure, P. S., Srivatsava, A. K., & Shyam Singh, (2001). Fertigation and drip irrigation in Nagpur mandarin. *South Indian Hort.* 49, 95–96.

36. Singh, S. D., J. P. Gupta and Panjab Singh, (1978). Water economy and saline water use by drip irrigation. *Agron. J.,* 70, 948–951.

37. Srinivasachari, M., Dhakshinamoorthy, M., & Arunachalam, G. (1998). Effect of paper factory effluent on soil available macronutrients and yield of rice. *Madras Agric. J.,* 85 (2), 564–566.

38. Udayasoorian, C., & Prabakaran, C. (2010). Effect of fertigation on leaf proline, soluble protein and enzymatic activity of banana. *EJEAFChe.* 9(8), 1404–1414.
39. Viswanatha, G. B., Ramachandrappa, B. K., & Nanjappa, H. V. (2000). Effect of drip irrigation and methods of planting on root and shoot biomass, Tesseling-silking interval, yield and economics of sweet corn. Mysore, *J. Agric. Sci.*, 34, 131–141.

PART III

MICRO IRRIGATION PRACTICES IN AGRICULTURAL CROPS

CHAPTER 7

PERFORMANCE OF PULSE DRIP IRRIGATED POTATO UNDER ORGANIC AGRICULTURE PRACTICES IN SANDY SOILS

ABDELRAOUF RAMADAN

CONTENTS

---

[a]This chapter is an edited version of *Abdelraouf Ramadan Eid Abdelghany, 2009. Study the performance of pulse drip irrigation in organic agriculture for potato crop in sandy soils. PhD Dissertation at Department of Agricultural Engineering, Faculty of Agriculture, Cairo University, Egypt.*

[b]In this chapter: *1 feddan = 0.42 hectares = 4200 m² = 1.038 acres = 24 kirats.* A feddan (Arabic) is a unit of area. It is used in Egypt, Sudan, and Syria. The feddan is not an SI unit and in Classical Arabic, the word means 'a yoke of oxen': implying the area of ground that can be tilled in a certain time. In Egypt the feddan is the only non-metric unit, which remained in use following the switch to the metric system. A feddan is divided into 24 Kirats (175 m²). In Syria, the feddan ranges from 2295 square meters (m²) to 3443 square meters (m²).

[c]*One L.E. = 0.14 US$.* The Egyptian pound (Arabic: Geneh Masri-EGP) is the currency of Egypt. It is divided into 100 piastres, or (Arabic: 100 kersh), or 1,000 Millimes (Arabic: Millime). The ISO 4217 code is EGP. Locally, the abbreviation LE or L.E., which stands for (Egyptian pound) is frequently used. E£ and £E are rarely used. The name Geneh is derived from the Guinea coin, which had almost the same value of 100 piastres at the end of the 19th century.

7.1 INTRODUCTION

Pulse irrigation is used throughout the world because of its positive effects on yield, fruit quality, water saving, less clogging, and reduction in consumption of energy. Pulse irrigation refers to the practice of irrigating for a short period, then waiting for another short period, and repeating this on-off cycle until the entire irrigation water is completed [28]. Drip irrigation, nowadays, is most efficient plant watering system [37]. For efficient use of water, drip irrigation and pulse technique can be combined.

Pulse drip irrigation (PDI) has been used in combination with organic agriculture to get a major utilization from organic agriculture. Organic farming covers agriculture systems that implement the environmentally, socially and economically sound production of food and fibers. The production of organic agriculture products without inputs of chemical pesticides and fertilizers has become a profitable area of farming as consumers become more concerned about possible effects of chemicals.

Potatoes are largest horticultural export crop in Egypt. Organic production of potatoes is growing in Egypt to take significant place in the European market and to attract consumers who are willing to pay high price for a healthy safe product. In most recent years, the *Euros united* has accounted for about 70–90% of Egyptian potato market. In 2004, the total value of potato exports to the Europe was about 65 million Euros, about 43.5% of Egypt's agricultural exports to the Euros united [12].

This research study discusses results on the performance of pulse drip irrigation under organic agriculture for: saving water, saving fertilizers, increasing yield of potato, increasing the energy use efficiency, improving potato quality, decreasing the costs and increasing income under Egyptian growing conditions. The performance parameters were:

i. Soil moisture distribution.
ii. Water application efficiency.
iii. Clogging ratio of emitters.
iv. Emission uniformity.
v. Potato yield.

   vi. Water use efficiency.
  vii. Energy use efficiency.
 viii. Fertilizer use efficiency.
   ix. Quality characteristics of potato tuber.
    x. Economic analysis.

7.2 LITERATURE REVIEW

7.2.1 PULSE IRRIGATION

Pulse irrigation involves providing small amounts of water during the morning to modify the microclimate and keep the plants photosynthesizing longer. The main irrigation is applied during the early evening. This approach is used in countries such as Israel where crops such as avocado are grown in 'stressful' environments [17, 28, 65].

How often to irrigate (irrigation frequency)? The irrigation frequencies used in drip irrigation are typically quite different from those used in other methods of irrigation. The drip systems allow applying small amounts of water daily or several times a week without significant loss to evaporation or surface runoff. This ability to use frequent irrigation to keep the soil moisture level near field capacity is a unique advantage of drip irrigation. During consecutive days of hot dry weather, or when young seedlings are grown in coarse textured soil, daily irrigation is good practice to ensure that plants are not stressed. Irrigating several times daily may result in reduced distribution uniformity, since the repeated filling and draining of sub main and laterals with each irrigation results in heavier irrigation at low points of the field. As described above, "pulse irrigation" can help in the development of an adequate wetting pattern [65].

In recent years, scientists and engineers of the All-Russian Research Institute of Irrigation Systems and Rural Water Supply (RADUGA) under the Ministry of Agriculture and Food of the Russian Federation in collaboration with scientists of some institutes of the country have carried out theoretical and experimental studies, pilot elaboration and development of new technologies and equipment for pulse micro irrigation. The efficiency of synchronous pulse sprinkling has also been studied in Germany, Cuba, Poland, Puerto Rico, India, and Bulgaria. The technologies of localized pulse irrigation were studied under greenhouse conditions in vegetable crops in Moscow and the Krasnodar Territory. The technologies of fertilizer application according to plant requirements have been worked out in tea plantations of the Krasnodar Territory and in vegetable plantations of the Ivanovo and Moscow regions [52, 60].

Characteristics of pulse irrigation (for sprinkler, micro irrigation, localized trickle irrigation, mist irrigation) in Russia have been described along with water and fertilizer application technologies. Special emphasis was given to the agro-biological efficiency of systems and technologies. The effects of pulse Micro irrigation and localized irrigation technologies on yields of different crops (tea, orchards, beetroots, grass and grain crops) in some regions of Russia have been studied by Kolganov [51].

7.2.1.1 Effects of Pulse Drip Irrigation (PDI) on Wetted Zone

Approximately 70% of water used by plants is removed from the upper half of the plant root zone (Figure 7.1). Optimum crop yields result, when soil-water tensions in the root zone are kept below 5 atmospheres. Root penetration can be extremely limited due to dry soil, a water table, bedrock, and high salt concentration zones [27].

To schedule irrigation, amount of water available in the crop root zone is compared with the tree's daily water requirement. If the daily water requirement exceeds the amount of water that can be held in the root zone, one will

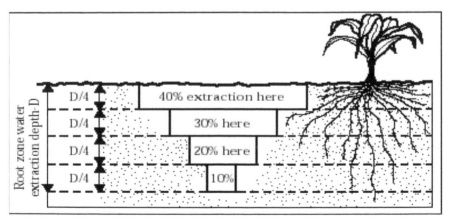

FIGURE 7.1 Amount of water uptake by roots in the upper half of the plant root zone.

need to irrigate more than once a day. There is an option of irrigating when the available water is depleted (called a deficit irrigation) [40]. Redesign of the irrigation system is necessary if the wetted area is too small (limiting) and pulsing is not an option [40]. Drip irrigation permits greater control and precision of irrigation timing and the amount of water applied [70]. Most growers scheduled a pulse irrigation when soil moisture tension in the main part of the root zone (15–20 cm) reached about −10 cbars [75].

Studies in Israel indicate that on Israeli sandy loam soils, pulse irrigated citrus had the main root zone within the first 30 centimeters [36]. Based on reports from other regions (where soil types are different), it is often believed that the size of the wetted zone can be increased if irrigation is pulsed [28].

The results showed a consistent trend that pulsing irrigation reduce the vapor pressure deficit about the trees. As increasing vapor pressure deficit can increase the plants internal water deficits, it can be assumed that the pulsing irrigation regime was effective in reducing water stress within the tree. Information from the USA and Israel provides some evidence that the *Open Hydroponic Principle* of maintaining soil moisture as close as possible to field capacity can provide productivity and water saving improvements. This form of irrigation management includes pulse irrigation and low output continuous irrigation during the day. Research on sandy soils in Florida indicates that trees begin to experience water stress when soil moisture levels fall below field capacity [10].

For most crops, the soil in the root zone should be kept near field capacity at all times. This means that irrigation should be frequent, and the amount of water applied each time should be equal to the amount used by the plants since the last irrigation. In general, short irrigation cycles with high application rates help promote lateral movement of water, resulting in better wetting patterns for light soils. Pulse irrigation can further widen the wetted pattern. Long duration at a low application rate results in better infiltration of water in heavy (high clay content) soils. The right irrigation cycle depends on the specifics of your field – experiment to find out what works best (Table 7.1).

The goal of drip irrigation scheduling is to select an irrigation duration and frequency that results in a properly sized wetted area around plants and keeps the soil in the root zone at or near field capacity. Adjustments

TABLE 7.1   Effects of Wetted Area on Crops

Small wetted area	Large wetted area
• Restricts roots to a small volume of soil.	• Wastes water and fertilizer.
• Reduces uptake of needed minor nutrients from soil.	• Increases the numbers of weeds.
• Increases potential for plant water stress during periods of high temperature and wind.	• Does not improve crop performance.

Ideal wetted area

• The ideal wetted area is shaped as shown here.

• The wetted area should be maintained at the same size throughout the season to prevent salts near the edges from damaging the crops.

• Soil type and field preparation affect the shape of the cross section dramatically.

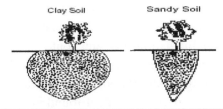

throughout the season based on monitoring of field conditions allow one to fine-tune the irrigation schedule to the needs of your crop [65].

Irrigation frequency is one of the most important variables in drip irrigation scheduling. Due to the differences in soil moisture and wetting pattern, crop yields may be different when the same quantity of water is applied under different irrigation frequencies. High irrigation frequency might provide desirable conditions for water movement in soil and for uptake by roots [68]. Several experiments have shown positive responses in some crops to high frequency drip irrigation [31, 68, 69].

Drip irrigation only wets soil near the plants. Roots only develop in the wetted area. This normally does not cause problems, but it makes irrigation frequency critical. Because the water holding capacity in the root zone is smaller, an extended period of time without irrigating can easily cause plant stress. During hot weather conditions, daily irrigation may be necessary to avoid crop damage from water stress [65]. Pulse irrigation system, irrigating amount and timing are the objectives for reducing run off, decreasing percolation of water beneath the root zone and reducing water evaporation after irrigation [23].

El-Adi [21] reported that soil moisture content and wetted area were increased with surge irrigation compared with traditional irrigation. Soil moisture content was increased by increasing irrigation water level.

Applying irrigation water in stages or pulses rather than all at one time can save water by giving the media time to moisten from the first pulse of water thereby allowing it to absorb subsequent irrigation more readily and reducing the total amount of water required. For example, instead of irrigating 4 different areas for 1 hour each (four hours total), studies have shown that by watering each area sequentially for 15-minute intervals and repeating this process twice, a 25% reduction in water usage (3 hours total irrigation time) can be obtained [66]. High irrigation frequency might provide desirable conditions for water movement in soil and for uptake by roots [68].

Micro irrigation enables the increase in irrigation frequency from weeks to daily or even shorter time periods, and enables management of soil water. The spatial fluctuations in water content enable plants to extract water from zones where water content is higher than field capacity, without the growth inhibiting effects of poor aeration. Augmented mass flow in the soil caused by the high water content found in high-frequency irrigation regimes increases water availability due to higher water potential and hydraulic conductivity. High-frequency irrigation resulted in greater water consumption and greater yields of sunflower under lysimeter conditions. Continuous irrigation resulted in even greater growth than irrigation consisting of 8 pulses per day [68].

One of the most common problems encountered with drip irrigation is that irrigation is applied in excess of crop requirements, whereby the water saving potential of drip is forfeited. According to Bravdo and Proebsting [11], the water distribution pattern under a dripper normally forms a bulb- or onion-shaped zone [37]. If over irrigation occurs, the bulb-shaped wetted zone gradually becomes carrot-shaped and eventually may form a "chimney" leading the excess water downward towards the water table [43]. Valuable nutrients may be leached out of the root zone and become unavailable for the plants, while contaminating the groundwater. Temporary aerobic conditions may then occur in the rooting zone, hampering growth and development and eventually yield and quality of tubers. According to Van Loon [81], the depth and horizontal water flow can be manipulated by changing the frequency of irrigation.

7.2.1.2 Effects of Pulse Drip Irrigation on Fertilizer Use Efficiency (FUE)

The benefits of pulse irrigation method are: reducing surface soil water evaporation, decreasing fertilizers leaching, enhancing yield [85]. The design modules have been elaborated with regard to the technologies of pulse low-volume application of water and fertilizers in accordance with plant requirements: pulse sprinkler micro irrigation with self-oscillating sprinklers and sprinklers operating in a "waiting mode"; pulse mist irrigation with pneumatic hydraulic sprayers to be mainly used under greenhouse condition; pulse trickle localized irrigation for greenhouse and open-ground conditions [50].

Using ebb-and-flood or pulse irrigation, fertilizer amounts can be reduced to avoid high electrical conductivity build up in the media due to reduced leaching. However, leaching is recommended from time to time to avoid salt buildup and crop damage. Leaching should be based on the electrical conductivity measurements of the soil solution [53]. The advantages of pulsing are that plant growth is generally greater than with standard irrigation and lower fertilizer rates can be used [16].

7.2.1.3 Effects of Pulse Drip Irrigation on Clogging Ratio of Emitters

Pulse trickle localized irrigation systems can be used for vegetables under greenhouse conditions. It is possible to use these systems under open-ground conditions on relatively small land plots. These systems permit water application automatically, continuously, with low intensity and in a cyclic mode. Water is delivered to the sites at a relatively high discharge and for a short period of time. This mode of operation favors water application without thorough treatment of irrigation water. The probability of water outlets clogging is low, because their cross-sections exceed the sizes of particles, which remain in water after pretreatment. The distinctive feature of pulse irrigation systems with cyclic water delivery makes it possible to put into practice the technology of dosed application of dissolved fertilizers during each cycle of water delivery. This noticeably simplifies the mechanisms of dosing and operation of fertigators. The technique of dissolved fertilizer application within the systems of pulse irrigation should involve a container for fertilizer solution (mother liquor) and a device for pulse dosed delivery of this solution into an irrigation network [37]. The alternatives of centralized (at the head of the system) and decentralized (immediately at sprinklers) patterns of fertilizer delivery into an irrigation network are possible. The degree of automation of technological processes of water and fertilizer application within pulse micro irrigation systems with different design modules depends on the use of supplementary technological equipment and automation facilities [51].

In recent years, farms of Daghestan have gained positive experience in growing row crops (tomatoes) under drip irrigation, the equipment being supplied by the firms of Israel. At present, micro irrigation practices in Russia are limited by the following socioeconomic and technical factors: high capital investments in construction of micro irrigation systems, including water treatment, and insufficient reliability of structural elements of these systems; high disparity in prices of crop production, on the one hand, and expensive industrial equipment, on the other hand; increased demand of the used equipment for pure water; relatively low general level of irrigation farming and lack of confidence in new technologies of micro irrigation; clogging of micro-outlets for water. The main trends in development and use of micro irrigation equipment are as follows: increase in ability of operation of micro-outlets, provision of the high degree of water distribution within irrigation systems in time and space using simple devices. As a consequence, there is a reduction of capital investments in construction of systems; optimization of parameters and regimes of micro irrigation systems, which provide environmental safety and high economically sustainable crop yields. To a great extent, these requirements are met by the technologies of pulse micro irrigation developed in Russia [51].

Drip tape can become non-uniform to a point where it is completely debilitated in the midst of a growing season if emitters become plugged, due to: (1) Organic or inorganic sediment in the irrigation water. (2) A vacuum condition inside of the drip tape causing dirt to siphon back in through the outlet. (3) Root intrusion. (4) Mineral buildup in the flow channel or at the outlet. The primary features of an emitter that determine its likelihood of plugging are the cross-sectional area of its flow channel and the amount of turbulence created within the flow channel. A large cross-section gives plenty of room for contaminants to pass through without accumulating into clogs. A highly turbulent channel keeps dirt particles suspended as they move through the emitter. Other emitter features also play important role in plugging resistance. Some drip tape products have emitter outlets that resist root intrusion. The design of the emitter inlet can also affect clog resistance. Finally, some emitters provide mechanisms that help to remove clogs if they should occur [73]. Pulsed water flow can increase resistance to blockages in self-regulated trickle irrigation and avoid costly and bulky filters. It ensures more uniform irrigation over the whole cycle [83].

One effective means of avoiding emitter clogging in drip irrigation is to use larger emitter orifices. However, to maintain the application rate at optimum, the system has to be operated in pulses rather than continuously [4]. A laboratory study on a sandy loam soil demonstrated that pulsed flows up to three times the equivalent continuous flow can be used with little change to the soil wetting pattern, enabling a significant increase in emitter sizes and a reduced tendency to clog [44].

7.2.1.4   Effects of Pulse Irrigation on Water Application Efficiency (WAE)

It is important that the system be well-designed hydraulically to assure high uniformity and distribution of water efficiently. Efficiency of water application depends on the system, design, management skills, and irrigation scheduling. Irrigation application efficiency is the ratio of the volume of irrigation water stored in the root zone and available for crop use (evapotranspiration) to the volume delivered from the irrigation system. This ratio is always less than 1.0 because of losses due to evaporation, wind drift, and leaching, which may occur during irrigation. Improved irrigation efficiency can lead to reductions in water and energy consumption, more effective nutrient use and disease management, better yields, and improved crop quality and erosion control management systems [34].

The systems have undergone immense development in recent years and now allow the simple and accurate timing of irrigation events. The level of control includes the ability to "pulse" irrigation events to meet the needs of soils that have less than desirable infiltration rates, thus minimizing run off [80]. Bader et al. [5] concluded that the emission uniformity of drip irrigation system ranged from 81 to 83%, due to the effects of emitter discharge rate, which have been affected by emitter clogging.

Potential groundwater contamination can be reduced due to the decreased leaching of irrigation solution and reduced fertilizer concentrations. The pulse irrigation system virtually eliminates irrigation solution from leaching out of the pot. The irrigation solution is not leaching from the pot that means the solution fails to make it to the ground, thus decreasing the potential of contaminating the groundwater. The other advantage of reducing flow of solution through the soil profile is that other contaminates that exist in the ground may not get a chance to leach into groundwater sources. Instead of paying exorbitant amounts of money to reconstruct a facility to reduce groundwater contamination, greenhouse operations now have an alternate plan for reducing potential groundwater contamination for a lower cost. The operation must first obtain a time clock that can irrigate by seconds, not minutes like many controllers used in the industry do. Second, each individual working with the crops must learn how to produce a high quality crop with the pulse irrigation system. Recommendations for using the pulse irrigation strategy are to first lower fertility rates that are applied to the crops. Institute an aggressive and consistent soluble salt monitoring program, especially for crops that are sensitive to high soluble salt levels. Soluble salt levels can rapidly increase throughout the crop cycle due to the reduction of soluble salt leaching. Once individuals learn how to grow with the pulse irrigation system, a marketable crop can be produced that is similar to the quality they produced before using the pulse strategy [67].

Continuous water application is associated with increased water percolation under root zone. Intermittent irrigation strategy based on discharge pulses followed by breaks could improve water management in the field and increase irrigation efficiency [61].

7.2.1.5   Effects of Pulse Drip Irrigation on Energy Consumption

A low-pressure system for pulse trickle irrigation on small farms is characterized by a comparatively low opening head (0.2–1.5 m) together with lower power requirements than conventional trickle irrigation systems [32].

Pulse drip irrigation systems, which realize substantial savings of water and energy, are regarded in Bulgaria as an efficient means of irrigating small plots, home yards and greenhouses from limited water supplies, without requiring an external energy source [33].

7.2.1.6 Effects of Pulse Drip Irrigation on Yield and Water Use Efficiency (WUE)

It is necessary to get maximum yield in agriculture by using available water in order to get maximum profit from per unit area because existing agricultural land and irrigation water are rapidly diminishing due to the rapid industrialization and urban development. Therefore, one needs to know and supply the right amount of water needed for the plants. Furthermore, it is essential to develop the most suitable irrigation scheduling to get optimum plant yield for different ecological regions as plant water consumption depends mostly on plant growth, soil and climatic conditions [29]. Higher yield with saving in water results in high WUE [37].

Pulse irrigation is used in countries such as Israel where crops such as avocado are grown in atmospherically 'stressful' environments. Pulse irrigation under these conditions results in larger fruit which has a major impact on farm profitability. The ability to produce fruit of the size required by markets such as the US and Canada will be vital to develop an export industry [15].

Feng-Xin et al. [30] evaluated effects of irrigation frequency on soil water distribution, potato root distribution, and potato tuber yield and water use efficiency during 2001 and 2002. They used six different drip irrigation frequencies: N1 (once every day), N2 (once every 2 days), N3 (once every 3 days), N4 (once every 4 days), N6 (once every 6 days) and N8 (once every 8 days), with total drip irrigation water equal for the different frequencies. The results indicated that drip irrigation frequency did affect soil water distribution, depending on potato growing stage, soil depth and distance from the emitter. Under treatment N1, soil matric potential (cm) Variations at depths of 70 and 90 cm showed a larger wetted soil range than was initially expected. Potato root growth was also affected by drip irrigation frequency to some extent: the higher the frequency, the higher was the root length density in 0–60 cm soil layer and the lower was the root length density in 0–10 cm soil layer. On the other hand, potato roots were not limited in wetted soil volume even when the crop was irrigated at the highest frequency. High frequency irrigation enhanced potato tuber growth and water use efficiency. Reducing irrigation frequency from N1 to N8 resulted in significant yield reductions by 33.4 and 29.1% in 2001 and 2002, respectively [30].

Total potato tuber yield was highest for scheduling based on soil water balance. Irrigation frequency influenced yield differently for the different scheduling methods. Tuber relative density was improved by pulse irrigation [76].

The effects on crop yield of drip-irrigation frequencies of two irrigations per day (2/day), one irrigation per day (1/day), two irrigations per week (2/week), and one irrigation per week (1/week) was investigated for lettuce (Lactuca sativa), pepper (Capsicum annuum), and onion (Allium cepa) grown on sandy loam and processing tomato (Lycopersicon esculentum) grown on silt loam during experiments during 1994 to 1997. All treatments of a particular crop received the same amount of irrigation water per week. Results showed that the one/week frequency should be avoided for the shallow rooted crops in sandy soil. Irrigation frequency had little effect on yield of tomato, a relatively deep-rooted crop. These results suggest that drip irrigation frequencies of 1/day or 2/week are appropriate in medium to fine texture soils for the soil and climate of the project site. There was no yield benefit of multiple irrigations per day [39].

Kolganov and Nosenko [51] present brief information on the types of systems and technologies of micro irrigation used in Russia, along with the analysis of the available experience in micro irrigation practices. It offers the data on the peculiarities of systems and design modules of pulse irrigation (sprinkler micro irrigation, localized trickle irrigation, mist irrigation) as well as the data on the specific features of water and fertilizer application technology synchronous with plant requirements. Special emphasis was given to the agro-biological efficiency of systems and technologies of pulse micro irrigation. Empirical and field test results of studying the impact of pulse micro irrigation and localized irrigation technologies on yields of different crops (tea, orchards, beetroots, grass, and grain crops) in some regions of Russia are presented in the chapter by Kolganov and Nosenko [51].

Segal et al. [68] stated that high frequency and low flow irrigation can increase WUE and yield by providing favorable conditions for

water movement in soil and for uptake by roots. As soil water conditions become more constant, plants are able to utilize water and increase
production. Further study is necessary to evaluate these finding on other
crops and to develop economically feasible methods for low discharge
and high frequency irrigation.

In micro irrigation, the water application rate is adjusted to plant water
demands. High-frequency pulse application was compared with conventional drip irrigation in Israel. The two methods were applied to avocado
and citrus orchards and to carnation plants in greenhouse experiments. In the
orchard plots, no runoff was generated using high-frequency pulse application, saving 40% of water. In the greenhouse experiment, the total amount of
water applied was 18,680 m^3/ha for conventional irrigation and 9,970 m^3/ha
for high-frequency pulse application. The reduced amounts of water applied
did not change the soil moisture significantly. Yields for both treatments were
similar [47]. Beeson [8] also stated that total canopy dry weight was greater
with pulse irrigation in *Elaeagnus pungens, Ligustrum japonica,* and *Photinia
X fraseri*. Research with pulse watering and lower fertilizer rates may allow
micro-tubes to be more water efficient and produce less runoff [84].

With low-volume pulse water application close to the evapotranspiration, the crop yields (clover, rye grass, timothy, fescue, tip onion, lettuce) were usually 1.3–2.5 times higher than the crop yields grown
under traditional irrigation method. Similar results were obtained at the
Russian Research Institute of Irrigation Systems and Rural Water Supply
(RADUGA) during the experiments on plants in early periods of development. Field experiments in some regions of Russia proved the possibility
of stable increment in yields of crops grown under the alternatives with
a high degree of compliance of water application rates with evapotranspiration: 35% increment for tea, 15–30% increment for fruits and berries,
30–50% for sugar beet, 37% for fodder beet [60].

7.2.1.7   Effects of Pulse Drip Irrigation on the Quality of Crops

Effect of different irrigation scheduling techniques and drip irrigation
frequencies were studied in potato on water use, tuber yield and quality.
No significant differences in frying chip color were observed between any

of the treatments. Regarding tuber specific gravity, significant differences only occurred between irrigation frequencies. The tuber density in pulse method (average of the three scheduling methods) was significantly better than that of the non-pulse treatment (Figure 7.2). According to Bravdo and Poebsting [11], an even water supply in high frequency irrigation resulted in better fruit quality in orchards, due to a constant volume of wetted soil, which ensured that a certain portion of roots were continuously exposed to air at the interface between irrigated and non-irrigated soil. Tuber specific gravities were increased with increasing irrigation frequency, possibly because of a constant volume of wetted soil which was conducive to optimal gas exchange [77].

Pulse irrigation gave marketable plants at lower fertility levels and the quality of pulse irrigated plants were similar to that of 10% leaching and ebb-and-flood irrigated plants. This enabled growers to efficiently use water resources and apply lower fertilizer concentrations while reducing potential groundwater contamination. Growers cannot use high rates of fertilizer with the pulse irrigation strategy due to the rapid increase of soluble salts (EC). Once growers learn how to grow with this new system, a marketable crop can be produced that is similar in quality to that produced before using the pulse strategy [67].

Regular soil samples should be analyzed to track the soluble salt level. When using the pulse strategy, salt levels must be maintained at low levels especially for crops sensitive to high soluble salts such as Vinca or

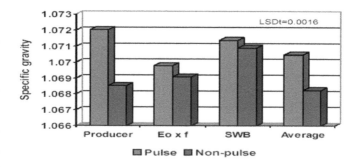

FIGURE 7.2   Potato tuber specific gravity for three scheduling methods at two irrigation frequencies.

New Guinea impatiens [16, 18]. Pulsing can be used with micro-tube and ebb-and-flood. With ebb-and-flood, most nutrient solution uptake was completed within the first 5 to 10 minutes of flooding [9]. Beeson [8] stated that pulse-irrigated plants tended to cause less daily water stress. With less water stress, plants grew faster and remained healthier than plants that were stressed on a daily basis. Also disease prevention was less difficult. Just as with ebb-and-flood and micro-tube strategies, foliage stays dry. Photonic leaf spot was shown to spread in the overhead irrigation system, while absent in the pulse-irrigated treatment.

The operation must first obtain a time clock that can irrigate by seconds, not minutes like many available controllers. Second, each individual working with the crops must learn how to produce a high quality crop with the pulse irrigation system. Recommendations for using the pulse irrigation strategy are to first lower fertility rates that are applied to the crops [67].

7.2.1.8 Automation of Irrigation

Rain bird [63] manufactured ESP-LX Modular controller, that is an irrigation timing system designed for commercial and residential use. The controller's modular design can accommodate from 8 to 32 valves. The controller includes many advanced features to help manage water efficiently. These include: (1) Programmable valve delay. (2) Cycle+Soak™. (3) Sensor connection with bypass switch. (4) Built-in diagnostic and validation software. (5) Compatibility with all Rain Bird remote systems, including one-button and multi-function systems.

The ESP-LX Modular controls when your drip system turns on, and how long the emitters run. The controller has several valves connected to it. Each valve opens when it receives power from the controller, and the emitters lines connected to that valve turn on. When these emitter lines have run for the programed time, the controller shuts off the valve and opens the next valve in sequence. For example, Figure 7.3 shows that valve 2 is currently watering. When valve 1 is finished, the controller will shut it off and start valve 2. In the same way, valve 3 will begin watering when valve 2 is finished.

Programming is the process of signaling the controller (Figure 7.4) exactly when and how long one wants to water. The controller opens and

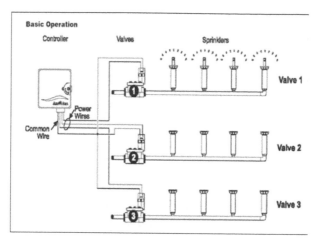

FIGURE 7.3   Layout of irrigation controller system.

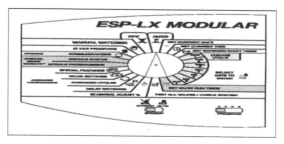

FIGURE 7.4   Keyboard of controller.

closes the remote control valves according to the setting on the keyboard. Each program contains: (1) Days to water: the specific days of the week on which watering takes place (e.g., Monday, Wednesday, and Friday), or the watering interval (e.g., every third day, or only on even or odd days of the month). (2) Watering start time(s): the time(s) of day that the program begins; this is the time that the first valve in the program begins watering; all other valves in the program then follow in sequence. (3) Valve run time: the number of minutes (or hours and minutes) that each valve runs.

Autonomic irrigation scheduling is a promising method to assist growers in conducting pulse irrigation. Although this technology is in its infancy and only applicable to small orchard situations, there is a lot of potential for its use in larger orchards. Controllers can be designed to accept from data electronic tensiometers and have inbuilt safety mechanisms to check if too much or too little water is being applied to the orchard. Some modular irrigation controllers already have the option of installing a special "PC card" that can conduct a form of autonomic irrigation control. Low output continuous irrigation should not be overlooked as a viable option to pulse irrigation. The principle is currently used by the *Martinez Open Hydroponic Technology* programs that design the irrigation systems to apply about 0.5 mm per hour. During summer, irrigation may occur for 10 hours per day during daylight hours. The method is an easy way to maintain soil moisture near field capacity. A recent desktop study demonstrated that pulse irrigation can maintain soil moisture closer to field capacity than continuous irrigation [36].

Selection of irrigation controller depends on:

1. Level of automation required.
2. Power source availability.
3. The number of stations to be irrigated.
4. Location of the controller.

For irrigation systems that have to be expanded over a period of time, then it may be more cost effective to install a controller with enough stations for the final system, or install a controller that allows for modular expansion. This also requires the need for provision for wiring for the final system. The basic function of a valve is to operate 'gates' to control the flow of water through the lateral lines. The two types of valves on the market are:

1. Manual – simple ball valve, gate valve and require an operator to activate the scheduling.
2. Automatic – hydraulic or electric hydraulic valves require a hydraulic command to open and close.

These valves use a small micro-tube (or a small diameter command tube) operating on the changes in water pressure through the command tube to open and close the valve. Electric valves require an electrical input from the controller to operate the valve. In-rush current is the power required to open a valve and the holding current is the power required to keep the valve in either the opened or closed position. Irrigation valves are normally closed, that is, require an inrush current to open and thus allowing water to flow through the valve. Latching valves only require a current to open or close the valve and do not require a holding current thus increasing wire run lengths. The valve configuration can be:

1. Angle: water enters through the base of the valve.
2. Globe: water enters the valve through the side.

The internal operation of the valve is hydraulic requiring venting of the water on or off the internal diaphragm to open or close the valve. Wiring for valves should be laid alongside the pipe work. The wiring size will depend upon the maximum run length between the controller and the valve. Each valve for a multi-core system will require a common solenoid. Black is the industry standard used for the common wire [59].

7.2.1.9   Economics of Pulse Drip Irrigation

A variety of computers and irrigation scheduling software is available. The cost of such a system can be a worthwhile investment. Install rain sensors to ensure irrigation does not occur during rain events. To lower costs and runoff, irrigate plants when needed based on media moisture content. This can be assessed by: (a) appearance or feel, (b) remote sensor tensiometers, (c) weight of media moisture, (d) light accumulators, and (e) moisture conductivity. Several devices relate media moisture to electrical conductivity. When the growing media dries to a preset level, the electronic circuit activates the solenoid valve [34].

The capital investments in the construction of irrigation systems of new type decrease by 20–60% in comparison with the systems of regular operation. This reduction is due to water flow distribution and minimization of diameters of the water conveyance network and also due to elimination of thorough water treatment necessary for drip irrigation systems. The empirical studies and optimization calculations have shown that the minimum discharge capacity and the minimum coefficient for the water conveyance network and equipment are provided by the system of continuous operation, when the degree of equipment use-in-time is equal to one. Energy consumption by pulse localized irrigation system is much lower than by drip irrigation system, because the required pressure for a design module is less than 3–5 mm of water column [51].

Salt movement in soil horizons and the mechanism of secondary salinization are dependent on moisture gradients and soil suction. Stable maintaining of the soil moisture level under new technologies without spasmodic changes inherent in the traditional irrigation prevents salt transport into upper soil horizons. The technologies of pulse fertigation are resource-saving. More productive use of natural precipitation in the zone of unsteady moistening contributes to the reduction of seasonal water application by 15–25% compared to the traditional technologies. The consumption of water and fertilizers per unit of crop yield decreases by 20–50% and 30–40%, respectively [51].

The new technologies of synchronous pulse irrigation are environmentally safe. Low intensity of water delivery, which approximates the intensity of evapotranspiration, eliminates the occurrence of the surface runoff and deep-water percolation in all soil types. This, in turn, prevents water erosion and secondary salinization, which are the main adverse effects of irrigation of lands under complicated soil and hydro-geological conditions as well as geomorphological conditions [51].

With the advent of new irrigation technologies, water management can become a less expensive task. With the new automatic irrigation systems, (ebb-and-flood, capillary mats, micro-tube, and pulse) irrigation requires less labor. Other benefits of these new systems are:

1. Uniform plants: All plants get very similar amounts of water.
2. Nutrient solution is recalculated or leach ate eliminated: Potential for groundwater pollution is eliminated. Growers who are using the system change the nutrient solution every few months. Therefore, less water is needed to grow a crop.
3. Flexibility: Pot sizes and spacing can be varied.
4. Adaptability: Systems can be used in most existing greenhouses and work well with fixed or movable benches.
5. Humidity: leaves remain dry. The dry bench surface also results in a lower humidity and increased temperature in the crop area [7].

Pulse irrigation offers one of the most economical alternatives when it comes to limiting runoff. In the future, more pulse irrigation research is needed to develop fertilizer recommendations and to investigate media-fertilizer interactions relevant to crop production [24].

7.2.1.10   Drip Irrigation System for Potato Crop

Mohamed and Abel-Rahman [57] studied the effects of surface drip irrigation, subsurface drip irrigation (SDI) and sprinkler irrigation with different soil moisture contents on potato production and WUE. The significant conclusions were:

1. Total quantity of water use under SDI was lower than the total quantity of water under surface drip irrigation and sprinkler irrigation.
2. Subsurface drip irrigation used less water consumption compared with surface drip irrigation and sprinkler irrigation.
3. Decreasing application efficiency with increasing the moisture content in the soil before irrigation; SDI was able to increase application efficiency compared with surface drip irrigation and sprinkler irrigation; maximum application efficiency was with surface drip irrigation at 0.25 moisture content of the field capacity before irrigation.
4. Increasing in WUE with increasing in moisture content before irrigation from 0.25 to 0.7 of field capacity, and maximum WUE was at 0.7 from field capacity.

7.2.1.11   Organic Agriculture and Importance of Potato Crop to Egypt

Organic agriculture covers agriculture systems that implement the environmentally, socially and economically sound production of food and fibers. The production of organic agriculture products without inputs of chemical pesticides and fertilizers has become a profitable area of farming as consumers become more concerned about possible health risks of chemicals [3].

Organic agriculture is developing rapidly, and statistical information is now available from 138 countries of the world. Its share of agricultural land and farms continues to grow in many countries. Figure 7.5 shows latest survey on organic farming worldwide [41]. The various issues on the benefits of organically produced food are [49]:

  a. **Food safety:** Organic agriculture reduces the risks of pathogens (zoonoses), mycotoxins, bacterial toxins and industrial toxic pollutants, compared to conventional agriculture. Reduced resistance to antibiotics in zoonotic pathogens indicates a better prognosis

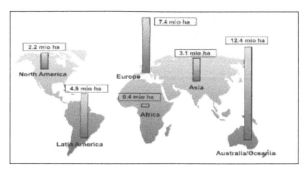

FIGURE 7.5 Organic farming worldwide.

for patients if an infection does occur. For natural plant toxins, the content in plants appears to systematically be 10–50% higher than in conventional plants.

b. **Pesticide poisoning:** This is an area where very substantial health problems have been documented, especially among farmers and their families. Pesticide poisoning causes some 20,000 deaths per year globally and an average of 11 days wages is lost due to illness, per farmer per incidence. Even symptom-free workers often exhibit biomarker changes indicating increased risk of diseases, including Parkinson's disease. With the present level of knowledge, elimination of such horrible conditions, which can be achieved on a short timescale, is the quantitatively single most important benefit of organic farming in terms of human health.

c. **Pesticide residues**: The levels in organic products are consistently 4 to 5 times lower than in conventional products. However, no definitive causal connection with harm to consumers has ever been demonstrated for food produced in accordance with general (conventional) food safety rules.

d. **Food quality**: Consumers generally appreciate that food is authentic and trustworthy and produced with care for them and the environment. So reduced food additives and pesticide residues, good traceability and emphasis on animal welfare all support the perception of organic food as being of high-quality.

e. **Nutritional adequacy**: In developing countries, organic agriculture has several advantages for the provision of nutrients, such as higher Zn/phytate ratio and better amino acid composition in cereals. Also, a more balanced diet due to the greater diversity of organic rotations, including legumes and various types of vegetables, and the need for animals on each farm provide important nutritional benefits. In developed countries, nutritional value is much more difficult to determine. However, the higher levels of plant secondary metabolites and conjugated fatty acids in milk may provide important protection against cardiovascular disease, cancer and other diseases known to be influenced by diet.

f. **Human health**: Epidemiological studies have shown better health scores among consumers of organic food for immunological characteristics and weight control, and similar benefits have been reproduced in animal studies, supporting a possible causal role of the food production system.

g. **Post-harvest operations**: Higher activity of plant defense mechanisms in organic plants reduces the losses during transport and storage.

h. **Pollution of drinking water**: Organic farmers have substantially higher economic incentives than conventional farmers to establish and maintain sufficient capacity for collection, composting and incorporation of animal and human wastes as valuable fertilizer. This is particularly important in areas where sanitation is not provided or standards are not enforced by the authorities. Such measures will also substantially reduce contamination with nitrates and phosphorus. There is little evidence that these minerals have any harmful effects on humans, if the drinking water is free of pathogens, except by promoting blooms of toxic algae.

i. **Pollution of the environment**: Organic agriculture protects the local environment against all types of pesticides (DDT) and has potential to benefit the global situation if the proportion of land under organic management becomes large enough to reduce the total use. Pollution with nitrate and phosphorus are major causes of eutrophication. Organic farms leach lower levels of phosphorus into drainage water than conventional ones. For nitrate, the loss from organic farms tends to be slightly lower than conventional, except when comparing organic outdoor pig production with conventional indoor production. However, recent data indicate that organically managed soil may be more efficient at denitrification, releasing most of the nitrate into the atmosphere as harmless N_2. If this is a general trend, the benefits of organic farming are much higher than mentioned here.

Certified Organic Agriculture started in Egypt 23 years ago in a small farm (SEKEM) of about 17 ha in the eastern desert to produce medicinal herbs for the export market. Expansion of this activity was quite slow until 1988. Thereafter a rapid growth occurred in the bio-dynamic production of vegetables, fruits, cereals, and cotton, beside the medicinal herbs. This rapid growth was initiated mainly by SEKEM and some other growers in Fayum and Kalubia governorates. In the fall of 1994, a new group of organic growers initiated the establishment of the *Union of Growers* and exporters of organic and bio-dynamic agriculture products. Members of this union produce and trade mainly in organic herbs, vegetables, fruits, potatoes, onion, and some cereals. Shortly after in the summer of 1998, a new organic project was started by Al-Hoda for agro-manufacturing due to the market demands for organic fruits and vegetables. In the meantime, Ever Green Egypt, Sonak, Sultan Farm, Fayum Society of Small Organic Farmers and others have been involved in the organic movement in Egypt. The expansion of organic agriculture activity in Egypt is growing very fast due to public awareness as well as the increasing demands for organic food and fibers in both local and export markets. The number of farms reached more than 500 with total acreage of more than 10,000 hectares. The total cultivated area in Egypt according to the Egyptian Ministry of Agriculture survey in the year 2000 is 3,083,333 hectares. Thus, organic farmed areas represent about 0.72% of the total area. Beside the certified Organic Agriculture production, there are more than 210,000 hectares that are cultivated traditionally without any use of chemicals in the remote areas, which depend on rain or underground water for irrigation [54].

By respecting the natural capacity of plants, animals and the landscape, organic agriculture aims to optimize quality in all aspects of agriculture and the environment. Organic agriculture dramatically reduces external inputs by refraining from the use of chemosynthetic fertilizers, pesticides, and pharmaceuticals. Instead, it allows the powerful laws of nature to increase both agricultural yields and disease resistance. Organic agriculture adheres to globally accepted principles, which are implemented within local social-economic, climatic and cultural settings. Organic agriculture may also be called ecological or biological agriculture or similar expressions in other languages than English [38].

Organic farming is based on the following approaches and production inputs: (1) Strict avoidance of synthetic fertilizers and synthetic pesticides. (2) Crop rotations, crop residues, mulches. (3) Animal manures and composts. (4) Cover crops and green manures. (5) Organic fertilizers and soil amendments. (6) Bio-stimulants, humates, and seaweeds. (7) Compost teas and herbal teas. (8) Marine, animal, and plant by-products. (9) Bio-rational, microbial, and botanical pesticides, and other natural pest control products [74].

In 1996, Egypt produced 2.6 million metric tons of potatoes and exported 411,000 metric tons valued at nearly US $80 million to Europe and the Arab countries. Small farmers grow 65% of these potatoes. In year 2000, Egypt produced 1.784 metric tons on an area of 83,000 ha with average yield of 21.49 tons per ha [26].

The potato is the 5th most important crop in the world. It is nutritious and highly productive, has a good value when sold, and is an effective cash crop for a developing country that has both local and export markets. This is the case in Egypt, where agriculture accounts for 28% of the national income. Almost 50% of the country's work force is dependent on the agricultural sub-sector. Rising population and the resulting increase in domestic demand for agricultural products are putting pressure on agricultural exports. However, the potato tuber moth, which mines the foliage and feeds on the tubers, is a serious pest of potatoes in both the field and in storage. Controlling the problem can increase Egypt's export volume by 15%: an additional $12 million in export income [14].

7.2.2 DEFICIT IRRIGATION

Over the past decade, Oregon State University Malheur Experiment Station at Ontario, Oregon, has evaluated drip irrigation on potato. The researchers investigated crop response to drip tape flow rate, bed conformation, and drip tape placement with respect to potato rows, micro irrigation criteria, and plant population. When compared to furrow irrigation, drip irrigation of potato reduces water use, nitrate leaching, erosion, and deep percolation, while increasing marketable yield. Drip irrigation of potato uses less water than sprinkler irrigation for comparable yield. Growers have many options for custom fitting a drip system to the specific situation. The publication by Shock et al. [71] provides a framework, general recommendations, and rationales to aid potato growers interested in maximizing their land use and crop yield through drip irrigation.

Deficit irrigation between jointing and anthesis significantly increased grain yield and WUE compared to rain fed treatment. The increased yield under irrigation was mainly contributed by increasing number of spikes, and seeds/m^2 and seeds/spike. The increased WUE under deficit irrigation was contributed by increased harvest index. When photosynthesis and biomass were reduced by water stress during grain filling, remobilization of pre-anthesis carbon reserves significantly contributed to the increased grain yield and harvest index. The irrigation timing is important for increasing yield and WUE under deficit irrigation [62].

Metin et al. [56] examined the effects of different irrigation regimes on yield and water use of bell pepper (*Capsicum annuum*) irrigated with a drip irrigation system under field conditions in 2002 and 2003 growing seasons in the Mediterranean region of Turkey. Irrigation regimes consisted of three irrigation intervals based on three levels of cumulative pan evaporation (Epan) values (I_1, 18–22 mm; I_2, 38–42 mm; and I_3, 58–62 mm). Irrigations occurred on the respective treatments when Epan reached target value. Three plant-pan coefficients were evaluated as irrigation levels ($Kcp_1 = 0.50$, $Kcp_2 = 0.75$ and $Kcp_3 = 1.00$). In conclusion, I_1Kcp_3 irrigation regime is recommended for field grown bell pepper in order to attain higher yield with improved quality.

Although the control of vegetative vigor in high-density orchards was the original objective of regulated deficit irrigation [35], yet increased WUE has become a critical issue in areas of water scarcity. Regulated deficit irrigation is an ideal water saving technique. Its application and adaptation in various environments have led to improved understanding of the process, the benefits, and the requirements for adoption. Scheduling has evolved to include weather and soil-based monitoring. As a consequence, this wealth of knowledge has enabled the implementation of a practical and achievable programmer for grower adoption of regulated deficit irrigation regulated deficit irrigation [35].

Agronomic measures such as varying tillage practices, mulching and anti-transpirants can reduce the demand for irrigation water. Another option is deficit irrigation, with plants exposed to certain levels of water stress during either a particular growth period or throughout the whole growth season, without significant reduction in yields [48].

Furthermore, yield reductions from disease and pests, losses during harvest and storage, and arising from insufficient applications of fertilizer are much greater than reductions in yields expected from deficit irrigation. On the other hand, deficit irrigation may increase crop quality. For example, the protein content and baking quality of wheat, the length and strength of cotton fibers, and the sucrose concentration of sugar beet and grape all increase under deficit irrigation [48].

Climate, agronomic and varietals factors determine total water use by a crop. Water deficit stress during the vegetative phase of development can increase WUE significantly. Soil moisture deficit stress during vegetative growth increased total biomass accumulation and pod yield, due to increases in leaf area during reproduction, and partitioning of more dry matter to the reproductive parts. In addition, the yield advantage due to water stress in the vegetative phase was due to improved synchrony in flowering and the increased peg-to-pod conversion. Moreover, stress during vegetative growth may have promoted root growth, an area which requires further study. The results show that it is possible to increase WUE and dry matter production, including the economic yield of groundnut crops cultivated under irrigated conditions by the imposition of transient soil moisture deficit stress during the vegetative phase. However, exact scheduling may differ in different environments [58].

Deficit irrigation can be used to increase the soluble solid content of fruits or vegetables by deliberately maintaining soil moisture below field capacity. This is usually done at the end of the growing season, shortly before harvest, and is common with grapes, sugar cane, tomatoes, cotton and several other crops. Precise control of application rates make drip irrigation ideally suited for deficit irrigation when necessary [65].

7.3 MATERIALS AND METHODS

7.3.1 CHEMICAL ANALYSIS AND PHYSICAL PROPERTIES OF SOIL, IRRIGATION WATER AND COMPOST

The soil excavation pit (Figure 7.6) was to obtain soil samples for determination of soil physical and chemical properties. The soil at the experimental site is sandy soil. Soil physical and chemical properties are presented in Table 7.2. Table 7.3 presents the analysis of compost (organic fertilizer) and irrigation water.

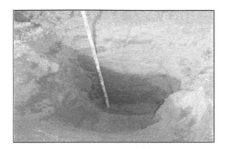

FIGURE 7.6 Soil excavation pit to obtain soil samples.

TABLE 7.2 Soil Physical Properties at Experimental Site

Property		Soil depth, (cm)			
		0–15	**15–30**	**30–45**	**45–60**
Particles Size distribution, (%)					
Very coarse sand, %		26.67	24.80	29.88	9.79
Coarse sand, %		21.70	23.52	24.71	15.82
Medium sand, %		24.80	30.64	30.40	24.86
Fine sand, %		19.10	15.96	12.37	41.60
Very fine sand, %		5.23	3.42	1.46	7.01
Silt + Clay, %		2.50	1.66	1.18	0.92
Soil texture		Sandy	Sandy	Sandy	Sandy
Soil physical property					
$CaCo_3$, (%)		8.5	8.5	8.0	7.6
Field capacity, (%)		10	10	10	9
Wilting point,(%)		3.6	3.5	3.4	3
Bulk density, g/cm³		1.49	1.55	1.59	1.61
Soil chemical property					
pH		7.43	7.60	7.75	7.79
EC, (ds/m)		3.72	3.50	3.20	2.87
	$HCO_3^- + Co_3^=$	2.40	1.20	1.50	1.25
Anions	CL-	23.0	22.5	18.1	14.5
(meq./L)	$SO4^=$	25.0	25.3	17.5	13.9

TABLE 7.2   *(Continued)*

	Ca^{++}	28.5	30.0	17.0	16.7
Cation	K$^+$	0.70	0.80	0.60	0.65
(meq./L)	Mg$^+$	8.5	8.0	7.0	5.0
	Na$^+$	12.7	10.2	12.5	7.3
Organic matter, (%)		1.36	1.70	1.35	2.00
Boron, (ppm)		2.0	2.1	2.4	2.5

TABLE 7.3   Chemical Analysis of Compost (Organic Fertilizer) and Irrigation Water

Property		Compost	Irrigation water
pH		5.92	7.55
EC, (ds/m)		0.70	1.90
Anions (meq./L)	HCO-$_3$ + Co$^=_3$	1.20	2.97
	CL-	3.50	3.94
	SO4$^=$	2.90	11.36
Cation (meq./L)	Ca^{++}	2.00	6.53
	K$^+$	2.20	0.22
	Mg$^+$	1.00	4.72
	Na$^+$	2.40	6.80
Organic matter, (%)		97.2	—
Moisture content, (%)		18.00	—
Nitrogen, (%)		0.91	—
Phosphorus, (%)		0.85	—
Potassium, (%)		0.90	—

7.3.2   DESCRIPTION OF THE IRRIGATION SYSTEM

The irrigation system consisted of: control head (pumping and filtration unit) as shown in Figure 7.7. It consisted of: submersible electrical pump with 45 m³/h discharge screen filter, back flow prevention device, pressure regulator, pressure gages, flow-meter, control valves. Main line was

FIGURE 7.7   Pumping and filtration units for drip irrigation system.

PVC pipes with 110 mm in diameter (OD) to convey the water from the source to the main control points in the field. Sub-main lines were PVC pipes with 75 mm diameter (OD) and were connected to the main line. Manifold lines consisted of PE pipes of 63 mm in diameter (OD) and were connected to the submain line through a 2″ control and discharge valves. Emitters were built-in with a dripper spacing of 30 cm on the laterals tubes of PE of 16 mm diameter (OD) and of 30 m in length. Emitter discharge was 2.1 lph at 1.0 bar of operating pressure.

7.3.3   EXPERIMENTAL SITE

During two summers of growing seasons in 2006 and 2007, the field experiment was carried out in Abo-Ghaleb farm, Cairo – Alexandria Rood, 60 Km away from Cairo. The soil is sandy. Seed tubers (var. Spunta Netherland) were planted on 2nd February and were harvested on 28th May in the two seasons.

7.3.4   FERTILIZATION METHOD AND BIO-FERTIGATION

Fertilizer requirements of potato crop were based on the recommendation of Field Crop Research Institute, ARC, Ministry of Agriculture and Land

Reclamation. The recommended dose of fertilizer for potato was 11 kg of Microbin mixed with one ton of seed tubers, and was directly before planting. The compost had 0.91% nitrogen and potato required from 150 kg of nitrogen. Therefore, total amount of compost was 150/9.1 = 16.48 tons/feddan and was applied each year. Microbin reduces it by 25% from applied compost. Hence, actual amount of compost for 2 years was = [(16.48 x 2) – 0.25(16.48 × 2)] = 24.72 tons/feddan. The compost was applied 20 days before planting and was added in middle of the row. Bio-fertigation (compost tea) was prepared by mixing 200 liters of water in 40 kg of compost and stored for 48 hours, and this mixture was injected through drip irrigation network two times weekly as shown in Figure 7.8. Table 7.4 shows physical properties at experimental site after adding compost.

7.3.5   EXPERIMENTAL DESIGN

The experimental design was a split-split plot with three replications. Irrigation systems and water regime treatments were used as main plots and

FIGURE 7.8   Injection of the compost tea in drip irrigation network.

TABLE 7.4   Physical Properties of Experimental Site After Adding Compost

Soil layers, cm	(0–15)	(15–30)	(30–45)
Field Capacity, (%)	12.6	11.8	11.6
Bulk density, g/cm³	1.40	1.48	1.52

submain plots (Figure 7.9). The pulse irrigation treatments were used as sub-sub main plots. Two irrigation systems were selected to irrigate potato plants. The surface drip irrigation consisted of built-in drip lines system with 30 cm emitters spacing; and PE laterals with 16 mm diameter at 70 cm spacing. The subsurface drip irrigation (SSDI) consisted of built-in drip lines system with 30 cm emitters spacing; and PE laterals with 16 mm diameter at 70 cm spacing and at 15 cm depth below soil surface [22]. Three water application rates were used: 50, 75 and 100% of actual water requirements (WRa) for potato. Three levels for pulse irrigation (2 times per day, 3 times per day and 4 times per day and the time-off between pulses was 30 minute) were compared with continuous drip irrigation (one time per day).

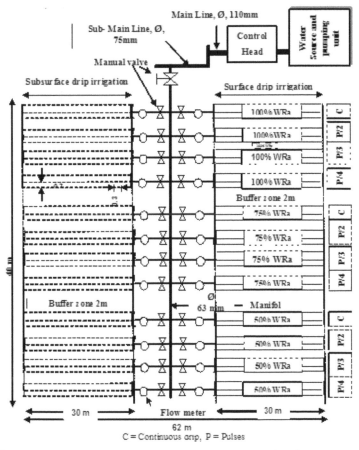

FIGURE 7.9 Layout of irrigation system with the experimental design.

7.3.6 ESTIMATION THE TOTAL IRRIGATION WATER

The total irrigation water for potato crop was estimated according to the meteorological data of the Central Laboratory for Agricultural Climate (CLAC) as shown in Figure 7.10 and Appendix I. It can be observed that the volume of applied water was increased with the plant growth, and then declined after the peak rate. The seasonal irrigation water applied was 3476 m³/fed. Tables 7.5 and 7.6 show the scheduling of pulse drip irrigation at maximum actual water requirements.

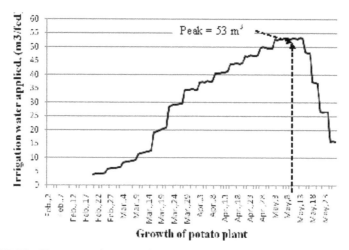

FIGURE 7.10   The potato plant growth versus and irrigation water application (m³/day).

TABLE 7.5   Scheduling of Pulse Drip Irrigation at Maximum Actual Water Requirements According to Experimental Design: One Pulse Per Day for Continuous Drip Irrigation (CID)

% from actual water requirements	Criteria	CID, one pulse/day	
		A.W.	O.T.
		m³	min.
	100	53	75.7
	75	39.75	56.8
	50	26.50	37.8
Time-off between pulses			0

TABLE 7.6 Scheduling of Pulse Drip Irrigation At Maximum Actual Water Requirements According to Experimental Design: 2, 3 and 4 Pulses/Day

Criteria	Pulse drip irrigation, pulses per day					
	Pulses per day					
	2		3		4	
	A.W.	O.T.	A.W.	O.T.	A.W.	O.T.
	m³	min.	m³	min.	m³	min.
100	26.5	37.8	17.7	25.2	13.3	18.9
	+	+	+	+	+	+
	26.5	37.8	17.7	25.2	13.3	18.9
			+	+	+	+
			17.7	25.2	13.3	18.9
					+	+
					13.3	18.9
75	19.8	28.4	13.3	18.9	9.9	14.2
	+	+	+	+	+	+
	19.8	28.4	13.3	18.9	9.9	14.2
			+	+	+	+
			13.3	18.9	9.9	14.2
					+	+
					9.9	14.2
50	13.3	18.9	8.8	12.6	6.6	9.4
	+	+	+	+	+	+
	13.3	18.9	8.8	12.6	6.6	9.4
			+	+	+	+
			8.8	12.6	6.6	9.4
					+	+
					6.6	9.4
Time-off between pulses	30		30		30	

% from actual water requirements

d = day, A.W. = Amount of water, CDI = Continuous drip irrigation, O.T. = Operating time.

7.3.7 DETERMINATION OF SOIL MOISTURE DISTRIBUTION

Soil moisture distribution was determined according to Liven and Van Rooyen [55]. The soil samples were taken at maximum actual water requirements by auger before and 2 hours after irrigation and from different locations. In case of 70 cm spacing between rows, the sample locations were taken at 0, 10, 20, 30 on the X- direction. In the case of 30 cm emitter spacing, the sample locations were at 5, 10, and 15 cm on the Z-direction (space between the emitters). For each of these locations, soil samples were collected from different depths from soil surface, which were 0, 15, 30 and 45 cm on the Y-direction. Soil moisture content was measured by the gravimetric methods. Soil moisture content percentage (S.M.W.) was determined on dry basis with the following equation:

$$S.M.W = [(W_1 - W_2) \times 100]/W_2 \qquad (1)$$

where: W_1 = weight of the wet soil sample (g), and W_2 = weight of the oven dried soil sample (g) at 105°C after 72 hours.

By using *contour program Surfer*, contour maps were plotted for different moisture levels at varying soil depths.

7.3.8 DETERMINATION OF APPLICATION EFFICIENCY

According to El-Meseery [25], application efficiency (AE) was calculated using the following equation:

$$AE = [V_s / V_a] \times 100 \qquad (2)$$

$$V_s = (\theta_1 - \theta_2) \times d \times \rho \times A \qquad (3)$$

where: AE = application efficiency (%); V_s = volume of stored water in root zone (cm^3) by calculating the wetted soil volume surrounded by contour line 12 that approximately represents the field capacity and the wetted soil volume surrounded by contour line 12 was converted to uniform volume; V_a = volume of applied water (cm^3); A = wetted surface area of root zone (cm²); d = soil layer depth (cm); θ_1 = soil moisture content after irrigation (%) at 15 m distance from lateral (at the middle of treatment); θ_2 = soil moisture content before irrigation (%) at 15.3 m distance from lateral; and ρ = relative bulk density of soil (dimensionless).

7.3.9 DETERMINATION OF CLOGGING RATIO OF EMITTERS

According to Sultan [79], the clogging ratio of emitters (CRE, %) was determined as follows:

$$CRE = 100 \times [1 - \eta], \text{ and } \eta = q_{used} / q_{new} \tag{4}$$

where: η = efficiency of emitter (fraction), q_{used} = average discharge for used emitters (lph); and q_{new} = Average discharge for new emitters (lph).

7.3.10 DETERMINATION OF EMISSION UNIFORMITY (EU)

Emission uniformity of drip irrigation system is a measure of the uniformity of emission from all the emission points within an entire drip irrigation system in the field. *EU* was determined according to the method by Keller and Karmeli [46]:

$$EU = (q_{min.}/q_a) \times 100 \tag{5}$$

where: *EU* = field emission uniformity (%); $q_{min.}$ = average rate of emitter discharge readings of the lowest one-fourth of the field data (lph); and q_a = average discharge rate of the emitters checked in the field (lph).

7.3.11 PLANT GROWTH PARAMETERS

A random sample of plants was taken at 60, 80 and 100 day after planting (DAP) for determination of plant height, number of stems/plant, number of leaves/plant, fresh weight/plant, dry weight/plant, and leaves area/ plant [3].

7.3.12 POTATO YIELD

At the end growing season, potato yields (tons/fed.) for each treatment was calculated as follows: (1) measure the area to determine the yield; (2) collect potato for each treatment; (3) weigh the potatoes for each treatment in the marked area; and (4) calculate yield in ton/fed.

7.3.13 DETERMINATION OF WATER USE EFFICIENCY (WUE)

Water use efficiency is an indicator of effectiveness irrigation unit for increasing crop yield. WUE (kg/m^3) of potato yield was calculated according to James [45]:

$$WUE = [\text{Total yield (kg/fed.)}]/[\text{Total applied water (m}^3/\text{fed.)}] \quad (6)$$

7.3.14 DETERMINATION OF ENERGY USE EFFICIENCY (EUE)

Energy Use Efficiency (kg/kW.h)) was determined according to Abdel-Aal [2]:

$$EUE = [\text{Total yield (kg)}]/[\text{Energy consumed (kW.h)}] \quad (7)$$

$$\text{Energy consumed (kW. h)} = BP \times \text{operating hours of irrigation} \quad (8a)$$

$$BP = (Q \times TDH \times Y_W)/(E_i \times E_P \times 1000) \quad (8b)$$

where: BP = brake power (kW); Q = discharge (m^3/sec); TDH = total dynamic head (m); E_P = pump efficiency (%); Y_W = water specific weight (9.81 kN/m^3); and E_i = electric engine efficiency (%).

7.3.15 DETERMINATION OF FERTILIZER USE EFFICIENCY (FUE)

Fertilizer use efficiency for N, P, K was determined according to Barber [6]:

$$FUE = \text{yield (kg/fed.)}/\text{Fertilizer applied (kg/fed.) for N, P and K} \quad (9)$$

$$FUE_N \text{ (kg/kg of N)} = [\text{Yield (kg/fed.)}]/[\text{Nitrogen applied (kg-N/fed.)}] \quad (10)$$

$$\text{FUE}_K \text{ (kg/kg of K)} = [\text{Yield (kg/fed.)}]/$$
$$[\text{Potassium applied (kg-K/fed.)}] \qquad (11)$$

$$\text{FUE}_p \text{ (kg/kg of P)} = [\text{Yield (kg/fed.)}]/[\text{Phosphorous}$$
$$\text{applied (kg-P/fed.)}] \qquad (12)$$

7.3.16 QUALITY OF POTATO TUBERS

A random sample of 20 tubers were selected from each experimental plot, then washed, dried and cut into small pieces (Figure 7.11) for determination of dry matter content, specific density and total carbohydrates.

1. Determination of dry matter content (TDMC) was determined after drying in an oven at 65°C for 72 hours using the standard methods as illustrated by A.O.A.C. [1].
2. Determination of tuber specific density (TSD, gm/cm^3) was measured as follows [13]:

$$\text{TSD} = [(\text{Dry matter content} - 24.182)/211.04] + 1.0988 \qquad (13)$$

3. Determination of tuber total carbohydrates (TTC) were determined (as glucose) after acid hydrolysis and spectro-photometrically using phenol sulfuric acid regent according to Dubbois et al. [20].

FIGURE 7.11 Cutting the tubers into small pieces for determination of dry matter content, specific density and total carbohydrates.

7.3.17   ECONOMIC ANALYSIS

Table 7.7 shows some details of economic analysis by Rizk [64]. Net income was determined by the following equation:

Net income (NI) = Total income for output – Total costs for Inputs   (14)

7.3.18   COST ANALYSIS OF IRRIGATION, LE/FED

Partial cost was conducted to evaluate differences between tested variables according to Worth and Xin [82]. The total cost (LE/year) for each treatment was calculated per feddan (60 m × 70 m), and described in Eq. (15). The market price level of 2007 was used for equipment and installation. Authors used following procedure.

TABLE 7.7   Method for Calculation of Net Income for Potato Production Under All Treatments

All treatments	Irrigation systems
	Water regime treatments
	Number of pulses
List of inputs	Cost of irrigation, LE/fed.
	Cost of land preparation, LE/fed.
	Cost of tuber seeds, LE/fed.
	Cost of compost, LE/fed
	Cost of Microbin, LE/fed
	Cost of weed control, LE/fed.
	Cost of pest control, LE/fed.
	Cost of harvesting, LE/fed.
	Rent (on season), LE/fed.
	Total cost for inputs, LE/fed.
Output	Yield, ton/fed.
	Price, LE/ton.
	Total income for output, LE/fed.
Net income = list of outputs – list of inputs.	

1. **Annual fixed cost** (F: LE/year) for the irrigation system was calculated using Eq. (16). Depreciation cost was calculated using Eq. (17). Interest rate/year (I.R.) was assumed as 14%. Taxes and overhead ratio (T, LE/year) were taken as 1.5% of initial cost. Annual operating cost (O: LE/year) on the capital investment for the irrigation system was calculated with equation (19). Energy cost (E) was calculated with Eq. (20).

$$\text{Total annual cost (LE/year)} = \text{Annual fixed cost (F)}$$
$$+ \text{Operating cost (O)} \qquad (15)$$

$$F = D + I + T \qquad (16)$$

$$D = (\text{I.C.} - \text{E.C.})/\text{E.L} \qquad (17)$$

$$I = (\text{I.C.} + \text{E.C.}) \times \text{I.R.} /2 \qquad (18)$$

$$O = L + E + (\text{R\&M}) + \text{IS} \qquad (19)$$

$$E = \text{Energy consumed (kW.h)} \times \text{Energy unit price (LE/kW.h)} \qquad (20)$$

In Eqs. (15)–(20): D = depreciation rate (LE/year), I = interest (LE/year), T = taxes and overhead ratio (LE/year), I.C. = initial cost (LE/fed.) = irrigation network item price = (LE) × (item quantity per fed.), E.C. = price after the depreciation (LE), E.L. = expected life (year), L = labor cost (LE/year), E = Energy cost (LE/year), R&M = repair and maintenance cost (LE/year) = 3 % of initial cost, and IS = cost for installation of laterals (LE/year).

7.3.19 STATISTICAL ANALYSIS

The standard analysis of variance procedure by Snedecor and Cochran [72] for split-split plot design with three replications was used. All data were calculated from combined analysis for the two growing seasons of 2006 and 2007. The treatments were compared based on the least square difference test (L.S.D.) at 5% level of significance.

7.4   RESULTS AND DISCUSSION

7.4.1   EFFECT OF PULSE DRIP IRRIGATION ON SOIL MOISTURE DISTRIBUTION IN THE ROOT ZONE

Soil moisture distribution is the main factor in the evaluation process for pulse drip irrigation (PDI) performance; and is affected by amount of applied water and number of pulses (frequency of irrigation in one day). Figures 7.12–7.23 indicate the relationship between PDI and wetted soil volume (more than or equal to 100% of field capacity: $WSV_{\geq 100\%FC}$) in root zone and the effects of PDI on moisture content in root zone (MCRZ under 100%, 75% and 50% of actual water requirements (WRa).

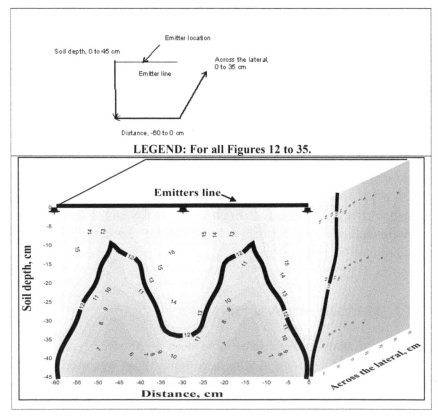

FIGURE 7.12   Three-dimensional soil moisture distribution and wetted soil volume (more than or equal 100% of FC) for sandy soil under surface drip irrigation at 100% of peak actual water requirements under continuous drip irrigation.

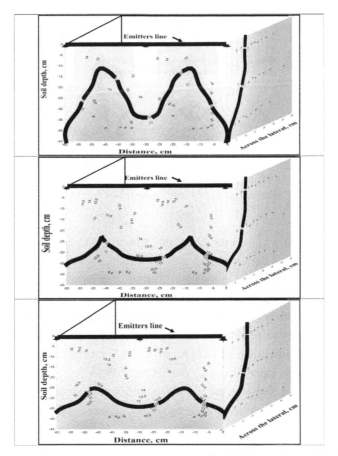

FIGURE 7.13 Three dimensional soil moisture distribution and wetted soil volume (more than or equal 100% of FC) for sandy soil under pulse surface drip irrigation (PSDI) at 100% of peak actual water requirements: (Top) on 2 pulses; (Center) on 3 pulses; and bottom) on 4 pulses.

7.4.1.1   PSDI on WSV$_{\geq 100\% FC}$ at 100% of WRa

WSV$_{\geq 100\% FC}$ in root zone was determined by calculating the wetted soil volume surrounded by contour line 12, which approximately represents the field capacity. Figures 7.12 and 7.13 show the relationship between pulse surface drip irrigation (PSDI) on WSV$_{\geq 100\% FC}$ at 100% of WRa. The WSV$_{\geq 100\% FC}$ in the root zone was increased by increasing number of irrigation pulses. Average of maximum width for contour line 12 from the emitter to 15 cm length was 8 cm and maximum depth was 45 cm, implying

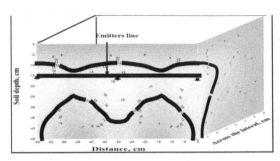

FIGURE 7.14 Three dimensional soil moisture distribution and wetted soil volume (more than or equal 100% of FC) for sandy soil under subsurface pulse drip irrigation at 100% of peak actual water requirements under continuous drip irrigation (CDI).

the area of $WSV_{\geq100\%FC}$ was 360 cm^2 and average of maximum width for contour line 12 cm from the emitter across lateral was 6 cm. Hence, $WSV_{\geq100\%FC}$ was 8640 cm^3 under continuous drip irrigation (CDI).

However, after applying the pulse technique with 4 pulses, the average of maximum width for contour line 12 from the emitter to 15 cm length was 15 cm and maximum depth was 31 cm, implying that the area of $WSV_{\geq100\%FC}$ was 465 cm^2 and average of maximum width for contour line 12 cm from the emitter across lateral was 12 cm. Hence, $WSV_{\geq100\%FC}$ was 22320 cm^3. $WSV_{\geq100\%FC}$ in root zone was increased from 8640 cm^3 under CDI to 22320 cm^3 under pulse technique with 4 pulses recording an increase of 61%.

7.4.1.2   PSSDI on $WSV_{\geq100\% FC}$ at 100% from WRa

Figures 7.14 and 7.15 indicate the relationship between pulse subsurface drip irrigation (PSSDI) on $WSV_{\geq100\% FC}$ at 100% from WRa. The $WSV_{\geq100\%FC}$ in the root zone was increased by increasing number of irrigation pulses. Average of maximum width for contour line 12 from the emitter to 15 cm length was 7.5 cm and maximum depth was 35 cm, this means that the area of $WSV_{\geq100\%FC}$ was 262.5 cm^2 and average of maximum width for contour line 12 cm from the emitter across lateral was 6 cm hence, $WSV_{\geq100\%FC}$ was 6300 cm^3 under CDI. However, after applying the pulse technique with 4 pulses the average of maximum width for contour line 12 from the emitter to 15 cm length was 15 and maximum depth was 22 cm, this means that the area of $WSV_{\geq100\%FC}$ was 330 cm^2 and average of maximum width for contour line 12 cm from

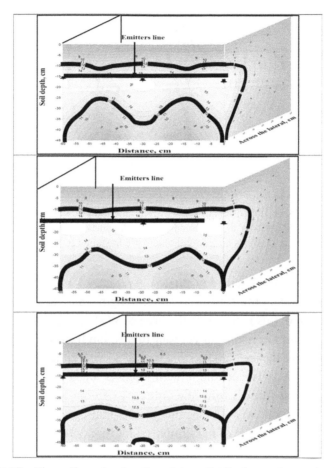

FIGURE 7.15 Three dimensional soil moisture distribution and wetted soil volume (more than or equal 100% of FC) for sandy soil under PSSDI at 100% of peak actual water requirements on: (TOP) 2 pulses, (CENTER) 3 pulses, and (BOTTOM) 4 pulses.

the emitter across lateral was 11 cm. Therefore, $WSV_{\geq100\%FC}$ was 14520 cm^3. $WSV_{\geq100\%FC}$ in root zone was increased from 6300 cm^3 under CDI to 14520 cm^3 under pulse technique with 4 pulses recording an increase of 57%.

7.4.1.3 PSDI on $WSV_{\geq100\%FC}$ at 75% from WRa

Figures 7.16 and 7.17 show the relationship between PDI on $WSV_{\geq100\%FC}$ at 75% from WRa. The $WSV_{\geq100\%FC}$ in the root zone was increased by

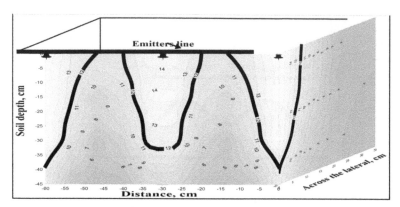

FIGURE 7.16   Three dimensional soil moisture distribution and wetted soil volume (more than or equal 100% of FC) for sandy soil under surface drip irrigation at 75% of peak actual water requirements under continuous drip irrigation (CDI).

increasing number of irrigation pulses. Average of maximum width for contour line 12 from the emitter to 15 cm length was 7 cm and maximum depth was 38 cm, this means that the area of $WSV_{\geq 100\%FC}$ was 266 cm² and average of maximum width for contour line 12 cm from the emitter across lateral was 6 cm. Hence, $WSV_{\geq 100\%FC}$ was 6384 cm³ under CDI. However, after applying the pulse technique with 4 pulses, the average of maximum width for contour line 12 from the emitter to 15 cm length was 10.5 cm and maximum depth was 38 cm, this means that the area of $WSV_{\geq 100\%FC}$ was 399 cm² and average of maximum width for contour line 12 cm from the emitter across lateral was 9 cm. Hence, $WSV_{\geq 100\%FC}$ was 14364 cm³. $WSV_{\geq 100\%FC}$ in root zone was increased from 6384 cm³ under CDI to 14364 cm³ under pulse technique with 4 pulses recording an increase of 56%.

7.4.1.4   PSSDI on $WSV_{\geq 100\%FC}$ at 75% from WRa

Figures 7.18 and 7.19 show the relationship between PSSDI on $WSV_{\geq 100\%FC}$ at 75% from WRa. The $WSV_{\geq 100\%FC}$ in the root zone was increased by increasing number of irrigation pulses. Average of maximum width for contour line 12 from the emitter to 15 cm length was 6 and maximum depth was 22 cm, this means that the area of $WSV_{\geq 100\%FC}$ was 132 cm² and average of maximum width for contour line 12 cm from

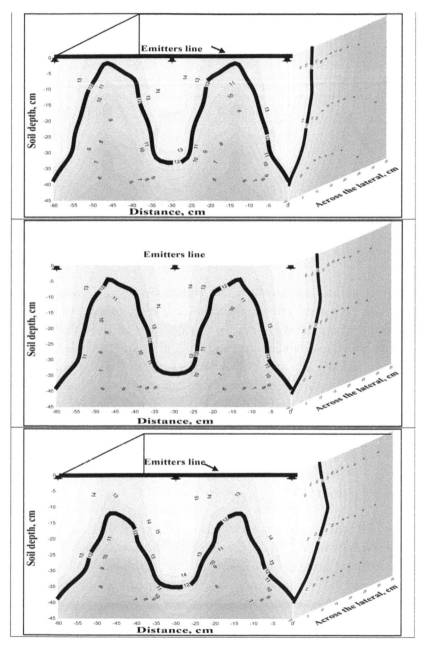

FIGURE 7.17 Three dimensional soil moisture distribution and wetted soil volume (more than or equal 100% of FC) for sandy soil under PSDI at 75% of peak actual water requirements on: (TOP) 2 pulses, (CENTER) 3 pulses, and (BOTTOM) 4 pulses.

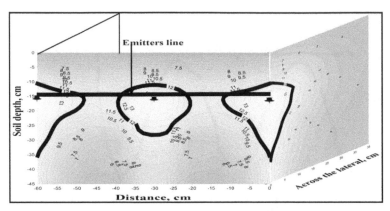

FIGURE 7.18   Three dimensional soil moisture distribution and wetted soil volume (more than or equal 100% of FC) for sandy soil under subsurface drip irrigation at 75% of peak actual water requirements under continuous drip irrigation (CDI).

the emitter across lateral was 6 cm. Hence, $WSV_{\geq 100\%FC}$ was 3168 cm^3 under CDI. However, after applying the pulse technique with 4 pulses, the average of maximum width for contour line 12 from the emitter to 15 cm length was 15 cm and maximum depth was 17 cm, this means that the area of $WSV_{\geq 100\%FC}$ was 255 cm^2 and average of maximum width for contour line 12 cm from the emitter across lateral was 10 cm. Therefore, $WSV_{\geq 100\%FC}$ was 10200 cm^3. The $WSV_{\geq 100\%FC}$ in root zone increased from 3168 cm^3 under CDI to 10200 cm^3 under pulse technique with 4 pulses recording an increase of 69%.

7.4.1.5   PSDI on $WSV_{\geq 100\% FC}$ at 50% of WRa

Figures 7.20 and 7.21 show the relationship between PSDI on $WSV_{\geq 100\% FC}$ at 50% of WRa. In $WSV_{\geq 100\% FC}$ in the root zone, there was increasing horizontal movement of water along contour lines by increasing number of pulses. $WSV_{\geq 100\%FC}$ in the root zone was too small under continuous drip irrigation CDI. However, after applying the pulse technique with 4 pulses, the average of maximum width for contour line 12 from the emitter to 15 cm length was 5 cm and maximum depth was 5 cm, this means that the area of $WSV_{\geq 100\%FC}$ was 25 cm^2 and average of maximum width for contour line 12 cm from the emitter across lateral was 4 cm. Hence, $WSV_{\geq 100\%FC}$ was 400 cm^3. $WSV_{\geq 100\%FC}$

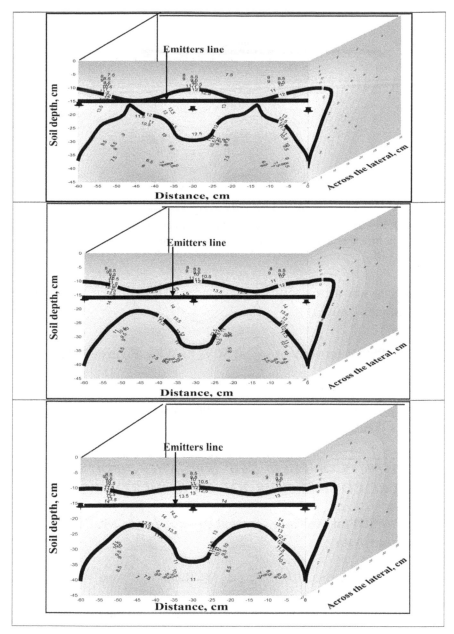

FIGURE 7.19   Three dimensional soil moisture distribution and wetted soil volume (more than or equal 100% of field capacity) for sandy soil under subsurface drip irrigation at 75% of peak actual water requirements on: (TOP) 2 pulses, (CENTER) 3 pulses, and (BOTTOM) 4 pulses.

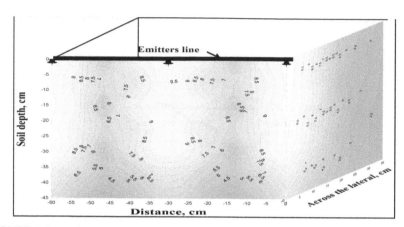

FIGURE 7.20   Three dimensional soil moisture distributions for sandy soil under pulse surface drip irrigation at 50% of peak actual water requirements under continuous drip irrigation (CDI).

in root zone was increased from point under CDI to 400 cm³ under pulse technique with 4 pulses.

7.4.1.6   PSSDI on WSV$_{\geq100\%\ FC}$ at 50% of WRa (Figures 7.22 and 7.23)

Figures 7.22 and 7.23 show the relationship between PSSDI on WSV$_{\geq100\%\ FC}$ at 50% of WRa.

7.4.2   EFFECTS OF PULSE DRIP IRRIGATION (PDI) ON SOIL MOISTURE CONTENT IN ROOT ZONE (MCRZ)

7.4.2.1   PSDI on Moisture Content in Root Zone "MCRZ" at 100% from WRa

Tables 7.8 and Figure 7.24 show that MCRZ in root zone was increased by increasing number of pulses. Average of moisture content was 12.18% in soil layer (0–15 cm), 10.43% in soil layer (15–30 cm), 7.53% in soil layer (30–45 cm) and MCRZ was 10.04%. Under CDI it was increased to 13.26% in soil layer (0–15 cm), 12.14% in soil layer (15–30 cm), 9.40% in soil layer (30–45 cm). The MCRZ was 11.60% after applying the pulse technique with 4 pulses recording an increase of 13.45%.

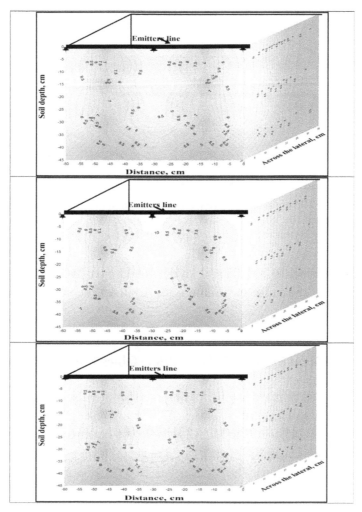

FIGURE 7.21   Three dimensional soil moisture distribution for sandy soil under PSDI at 50% of peak actual water requirements on: (TOP) 2 pulses, (CENTER), 3 pulses, and (TOP) 4 pulses.

7.4.2.2   PSSDI on MCRZ at 100% of WRa (Table 7.9 and Figure 7.25)

MCRZ was increased by increasing number of pulses. Average moisture content (MC) was 8.76% in soil layer (0–15 cm), 11.89% in soil layer (15–30 cm), 8.88% in soil layer (30–45 cm). MCRZ was 9.84% under CDI, but it changed to 9.39% in soil layer (0–15 cm), 12.45% in soil

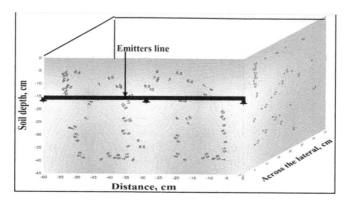

FIGURE 7.22 Three dimensional soil moisture distribution and wetted soil volume (more than or equal 100% of FC) for sandy soil under subsurface drip irrigation (PSSDI) at 50% of peak actual water requirements under continuous drip irrigation (CDI).

layer (15–30 cm), 10.30% in soil layer (30–45 cm). MCRZ was 10.71% after applying the pulse technique with 4 pulses recording an increase of 8.12%.

7.4.2.3 PSDI on MCRZ at 75% of WRa (Tables 7.10 and Figure 7.26)

The MCRZ was increased by increasing number of pulses. Average of MC was 10.81% in soil layer (0–15 cm), 9.27% in soil layer (15–30 cm), 6.95% in soil layer (30–45 cm). The MCRZ was 9.00% under CDI but it changed to 12.60% in soil layer (0–15 cm), 11.23% in soil layer (15–30 cm), 8.25% in soil layer (30–45 cm). The MCRZ was 10.69% after applying the pulse technique with 4 pulses recording an increase of 15.8%.

7.4.2.4 PSSDI on MCRZ at 75% of WRa (Tables 7.11 and Figure 7.27)

The MCRZ was increased by increasing number of pulses. Average of MC was 7.84% in soil layer (0–15 cm), 9.96% in soil layer (15–30 cm), 7.25% in soil layer (30–45 cm). The MCRZ was 8.35% under CDI but it changed to 8.86% in soil layer (0–15 cm), 11.50% in soil layer (15–30 cm), 8.93% in soil layer (30–45 cm). The MCRZ was 9.76% after applying the pulse technique with 4 pulses recording an increase of 14.44%.

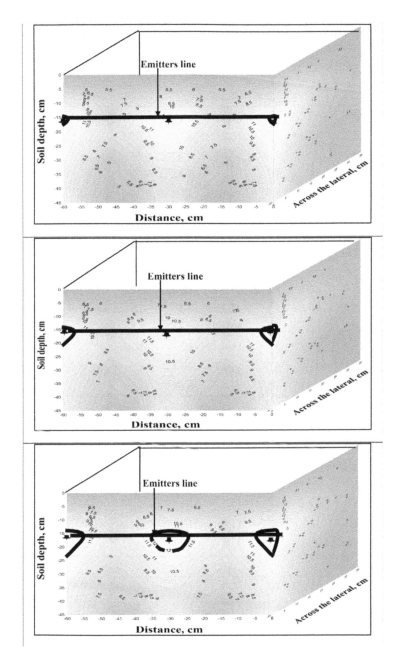

FIGURE 7.23 Three dimensional soil moisture distribution and wetted soil volume (more than or equal 100% of field capacity) for sandy soil under subsurface drip irrigation at 50% of peak actual water requirements on: (TOP) 2 pulses, (CENTER), 3 pulses, and (TOP) 4 pulses.

TABLE 7.8 Moisture Content in Root Zone Under Surface Drip Irrigation at 100% of Actual Water Requirements

Soil depth	Distance from emitter (across lateral)			Distance from emitter (along lateral)			Moisture content in soil layer	Moisture content in root zone
	0	10	20	5	10	15		
				Continuous drip irrigation (CDI)				
			cm				%	
0	16.6	12.1	5.0	16.2	12.1	13.0		
15	16.5	12.0	4.6	16.0	12.0	10.0		
30	16.0	8.0	4.0	12.0	8.00	6.00		
45	12.6	4.5	3.7	7.0	4.5	4.0		
0–15	16.6	12.1	4.8	16.1	12.1	11.5	12.18	
15–30	16.3	10.0	4.3	14.0	10.0	8.00	10.43	
30–45	14.3	6.25	3.9	9.5	6.3	5.0	7.53	10.04
PSDI, two pulses								
0	15.8	12.2	5.7	15.7	12.8	13.5		
15	15.7	12.1	5.6	15.6	12.3	10.5		
30	15.0	10.0	4.5	12.7	8.5	6.5		
45	12.5	6.3	3.7	6.5	4.8	4.5		
0–15	15.7	12.1	5.6	15.6	12.5	12.0	12.29	
15–30	15.3	11.0	5.0	14.1	10.4	8.5	10.75	
30–45	13.7	8.15	4.1	9.6	6.65	5.5	7.96	10.33
PSDI, three pulses								
0	16.0	12.5	6.6	15.8	13.0	13.6		
15	15.9	12.3	6.5	15.0	12.6	12.0		
30	13.0	11.5	5.0	13.0	12.0	10.0		
45	11.3	8.1	4.4	6.6	8.4	5.0		
0–15	15.9	12.4	6.5	15.4	12.8	12.8	12.65	
15–30	14.4	11.9	5.7	14.0	12.3	11.0	11.57	
30–45	12.1	9.8	4.7	9.80	10.2	7.5	9.03	11.08
PSDI, four pulses								
0	16.2	13.3	7.5	15.9	13.5	13.8		
15	16.0	13.0	8.5	15.5	12.9	13.0		
30	13.2	12.0	6.1	13.4	12.0	10.1		
45	11.4	8.20	5.4	7.00	8.60	5.40		
0–15	16.1	13.1	8.0	15.7	13.2	13.4	13.26	
15–30	14.6	12.5	7.3	14.4	12.4	11.5	12.14	
30–45	12.3	10.1	5.7	10.2	10.3	7.75	9,4	11.60

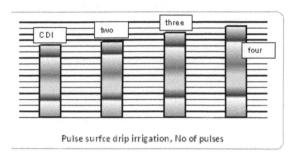

FIGURE 7.24 Effects of pulse surface drip irrigation on moisture content in root zone (MCRZ, Y-axis) under 100% of actual water requirements.

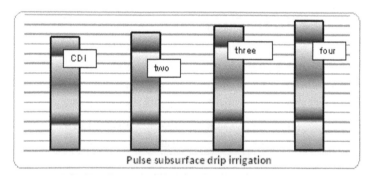

FIGURE 7.25 Effect of pulse subsurface drip irrigation on moisture content (MCRZ, Y-axis) in root zone under 100% of actual water requirements.

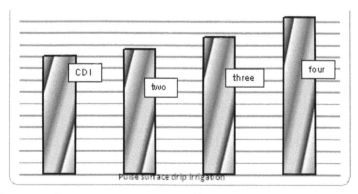

FIGURE 7.26 Effect of pulse surface drip irrigation on moisture content in root zone (MCRZ, Y-axis) under 75% of actual water requirements

TABLE 7.9 Moisture Content in Root Zone Under Subsurface Drip Irrigation at 100% of Actual Water Requirements

Soil depth	Distance from emitter (across lateral)			Distance from emitter (along lateral)			Moisture content in soil layer	Moisture content in root zone
	0	10	20	5	10	15		
	Continuous drip irrigation (CDI)							
	cm						%	
0	5.00	4.50	4.0	4.50	4.00	4.10		
15	16.0	12.1	8.0	15.9	15.0	12.0		
30	15.2	9.0	7.0	13.5	11.0	8.00		
45	13.6	5.20	5.7	9.00	5.40	4.00		
0–15	10.5	8.30	6.0	10.2	9.50	8.05	8.76	
15–30	15.6	10.5	7.5	14.7	13.0	10.0	11.89	
30–45	14.4	7.10	6.3	11.2	8.20	6.00	8.88	9.84
PSSDI, two pulses								
0	4.80	4.10	4.1	4.60	4.50	4.40		
15	16.0	12.3	8.3	16.0	15.5	12.50		
30	15.0	10.4	7.3	13.7	11.3	8.40		
45	12.5	6.00	5.7	9.40	5.50	4.20		
0–15	10.4	8.20	6.2	10.3	10.0	8.45	8.93	
15–30	15.5	11.3	7.8	14.8	13.4	10.45	12.23	
30–45	13.7	8.20	6.5	11.5	8.40	6.30	9.12	10.09
PSSDI, three pulses								
0	4.30	4.20	4.30	4.30	4.20	4.20		
15	15.5	13.8	9.0	15.4	15.4	13.4		
30	13.0	12.0	7.5	13.0	12.5	10.0		
45	12.0	7.00	5.8	11.0	7.00	6.50		
0–15	9.90	9.00	6.6	9.85	9.80	8.80	9.00	
15–30	14.2	12.9	8.2	14.2	13.9	11.7	12.54	
30–45	12.5	9.50	6.6	12.0	9.75	8.25	9.78	10.43
PSSDI, four pulses								
0	5.40	5.30	5.0	5.30	5.20	5.20		
15	14.5	13.0	11	14.4	14.4	14.0		
30	12.3	12.2	7.7	12.2	12.1	11.6		
45	12.0	8.40	5.9	12.0	9.00	8.20		
0–15	9.95	9.15	8.0	9.85	9.80	9.60	9.39	
15–30	13.4	12.6	9.3	13.3	13.2	12.8	12.45	
30–45	12.1	10.3	6.8	12.1	10.5	9.90	10.30	10.71

TABLE 7.10 Moisture Content in Root Zone Under Surface Drip Irrigation (PSDI) at 75% of Actual Water Requirements

Soil depth	Distance from emitter (across lateral)			Distance from emitter (along lateral)			Moisture content in soil layer	Moisture content in root zone
	0	10	20	5	10	15		
Continuous drip irrigation (CDI)								
	cm						%	
0	16.1	11.0	4.0	13.5	12.0	12.0		
15	16.0	10.4	4.0	13.0	10.6	7.00		
30	15.4	7.60	3.5	11.5	7.80	4.40		
45	11.5	4.40	3.2	6.00	4.60	3.50		
0–15	16.0	10.7	4.0	13.2	11.3	9.50	10.81	
15–30	15.7	9.00	3.7	12.2	9.20	5.70	9.27	
30–45	13.4	6.00	3.3	8.75	6.20	3.95	6.95	9.00
PSDI, two pulses								
0	15.8	11.2	4.7	13.7	12.4	12.20		
15	15.7	11.1	4.6	13.4	11.0	7.60		
30	15.4	8.7	3.8	11.7	7.90	4.50		
45	11.2	5.0	3.1	6.10	4.70	3.70		
0–15	15.7	11.1	4.6	13.5	11.7	9.90	11.12	
15–30	15.5	9.90	4.2	12.5	9.45	6.05	9.62	
30–45	13.3	6.85	3.4	8.90	6.30	4.10	7.15	9.29
PSDI, three pulses								
0	16.0	11.4	5.6	14.0	12.7	12.5		
15	15.8	12.2	5.8	13.5	11.5	7.90		
30	15.5	10.0	4.5	13.0	8.00	4.80		
45	11.0	7.10	3.4	6.50	4.90	4.00		
0–15	15.9	11.8	5.7	13.7	12.1	10.2	11.58	
15–30	15.6	11.1	5.1	13.2	9.75	6.35	10.21	
30–45	13.2	8.55	3.9	9.75	6.45	4.40	7.73	9.84
PSDI, four pulses								
0	16.3	11.5	5.7	16.0	13.0	13.4		
15	16.2	13.4	6.7	15.8	12.2	11		
30	15.0	10.2	5.0	14.0	8.30	7.00		
45	10.8	7.5	4.0	7.10	5.10	5.00		
0–15	16.2	12.4	6.2	15.9	12.6	12.2	12.60	
15–30	15.6	11.8	5.8	14.9	10.2	9.00	11.23	
30–45	12.9	8.85	4.5	10.5	6.70	6.00	8.25	10.69

TABLE 7.11 Moisture Content in Root Zone Under Subsurface Drip Irrigation at 75% of Actual Water Requirements

Soil depth	Distance from emitter (across lateral)			Distance from emitter (along lateral)			Moisture content in soil layer	Moisture content in root zone
	0	10	20	5	10	15		
Continuous drip irrigation (CDI)								
	cm						%	
0	4.20	4.10	3.5	4.20	4.10	4.30		
15	15.0	11.6	6.0	13.5	11.6	12.0		
30	13.6	7.00	4.5	10.3	7.00	7.40		
45	12.0	4.60	4.0	7.50	4.60	4.50		
0–15	9.60	7.85	4.7	8.85	7.85	8.15	7.84	
15–30	14.3	9.30	5.2	11.9	9.30	9.70	9.96	
30–45	12.8	5.80	4.2	8.90	5.80	5.95	7.25	8.35
PSSDI, two pulses								
0	4.30	4.20	3.70	4.30	4.20	4.50		
15	15.2	11.8	6.2	13.8	12.4	12.20		
30	13.8	8.00	4.6	10.6	7.40	7.60		
45	12.0	5.50	4.2	7.70	4.70	4.70		
0–15	9.75	8.00	4.9	9.05	8.30	8.35	8.07	
15–30	14.5	9.90	5.4	12.2	9.90	9.90	10.30	
30–45	12.9	6.75	4.4	9.15	6.05	6.15	7.57	8.64
PSSDI, three pulses								
0	4.30	4.20	3.7	4.30	4.30	5.00		
15	15.3	12.5	6.6	14.6	13.1	12.4		
30	13.9	8.80	5.5	12.5	8.10	8.00		
45	11.8	7.50	4.5	10.2	5.50	5.00		
0–15	9.80	8.35	5.1	9.45	8.70	8.70	8.36	
15–30	14.6	10.6	6.0	13.5	10.6	10.2	10.94	
30–45	12.8	8.15	5.0	11.3	6.80	6.50	8.44	9.25
PSSDI, four pulses								
0	4.80	4.70	4.3	4.80	4.50	5.00		
15	14.9	12.7	8.8	14.8	14.0	13.0		
30	13.1	10.5	6.6	12.9	8.50	8.20		
45	11.9	7.90	5.2	11.0	6.00	5.30		
0–15	9.85	8.70	6.5	9.80	9.25	9.00	8.86	
15–30	14.0	11.6	7.7	13.8	11.2	10.6	11.50	
30–45	12.5	9.20	5.9	11.9	7.25	6.75	8.93	9.76

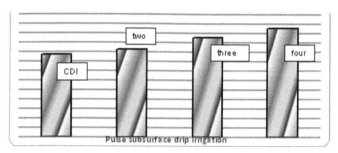

FIGURE 7.27 Effect of pulse subsurface drip irrigation on moisture content (MCRZ, Y-axis) in root zone under 75% from actual water requirements.

7.4.2.5 PSDI on MCRZ at 50% of WRa (Tables 7.12 and Figure 7.28)

The MCRZ was decreased by increasing number of pulses. Average of MC was 7.41% in soil layer (0–15 cm), 7.00% in soil layer (15–30 cm), 5.53% in soil layer (30–45 cm). MCRZ was 6.64% under CDI but it changed to 5.68% in soil layer (0–15 cm), 5.15% in soil layer (15–30 cm), 3.82% in soil layer (30–45 cm). The MCRZ was 4.88% after applying the pulse technique with 4 pulses recording decrease of 26.5%.

7.4.2.6 PSSDI on MCRZ at 50% of WRa (Tables 7.13 and Figure 7.29)

MCRZ was decreased by increasing number of pulses. Average of MC was 5.99% in soil layer (0–15 cm), 7.66% in soil layer (15–30 cm), 6.08% in soil layer (30–45 cm). The MCRZ was 6.58% under CDI but it changed to 5.78% in soil layer (0–15 cm), 7.28% in soil layer (15–30 cm), 5.65% in soil layer (30–45 cm) and MCRZ was 6.24% after applying the pulse technique with 4 pulses recording decrease of 14.28%.

Maximum $WSV_{\geq 100\% \ FC}$ was 22320 cm³ at 100% of WRa with 4 pulses under SDI and minimum $WSV_{\geq 100\% \ FC}$ was at 50% of WRa with CDI under SSDI. Maximum MCRZ was 11.60% at 100% of WRa with 4 pulses under SDI and minimum MCRZ was 4.88% at 50% of WRa with 4 pulses under surface drip irrigation. It is evident from the results in this section that there are significant differences in the soil moisture content in the root zone after applying pulse technique compared with continuous drip irrigation, because increasing number of pulses cause increase in water movement

TABLE 7.12 Moisture Content in Root Zone Under Surface Drip Irrigation at 50% of Actual Water Requirements

Soil depth	Distance from emitter (across lateral)			Distance from emitter (along lateral)			Moisture content in soil layer	Moisture content in root zone
	0	10	20	5	10	15		
				Continuous drip irrigation (CDI)				
			cm					%
0	10.30	9.00	3.80	9.50	9.00	3.70		
15	10.20	8.70	3.50	9.00	8.70	3.50		
30	10.00	7.60	3.40	8.80	7.60	3.00		
45	7.00	4.10	3.20	5.00	4.10	2.50		
0–15	10.25	8.85	3.65	9.25	8.85	3.60	7.41	
15–30	10.10	8.15	3.45	8.90	8.15	3.25	7.00	
30–45	8.50	5.85	3.30	6.90	5.85	2.75	5.53	6.64
PSDI, two pulses								
0	10.30	8.50	3.50	9.00	8.50	3.20		
15	10.00	8.20	3.00	8.50	8.20	3.00		
30	9.00	7.10	2.90	8.30	7.10	2.50		
45	8.00	3.60	2.70	4.50	3.60	2.00		
0–15	10.15	8.35	3.25	8.75	8.35	3.10	6.99	
15–30	9.50	7.65	2.95	8.40	7.65	2.75	6.48	
30–45	8.50	5.35	2.80	6.40	5.35	2.25	5.11	6.19
PSDI, three pulses								
0	10.20	7.90	2.90	8.40	7.90	2.40		
15	9.00	7.60	2.40	7.90	7.60	2.40		
30	8.00	6.50	2.30	7.50	6.50	2.00		
45	7.00	3.00	2.10	3.90	3.00	2.00		
0–15	9.60	7.75	2.65	8.15	7.75	2.40	6.38	
15–30	8.50	7.05	2.35	7.70	7.05	2.20	5.81	
30–45	7.50	4.75	2.20	5.70	4.75	2.00	4.48	5.56
PSDI, four pulses								
0	10.20	6.60	2.70	6.80	6.60	2.20		
15	9.00	6.20	2.40	6.80	6.20	2.40		
30	7.50	5.50	2.30	6.00	5.50	2.00		
45	6.00	2.00	2.00	3.00	2.00	2.00		
0–15	9.60	6.40	2.55	6.80	6.40	2.30	5.68	
15–30	8.25	5.85	2.35	6.40	5.85	2.20	5.15	
30–45	6.75	3.75	2.15	4.50	3.75	2.00	3.82	4.88

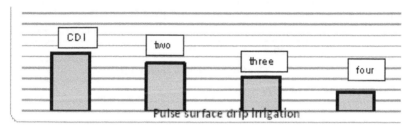

FIGURE 7.28 Effect of pulse surface drip irrigation on moisture content in root zone (MCRZ, Y-axis) under 50% from actual water requirements.

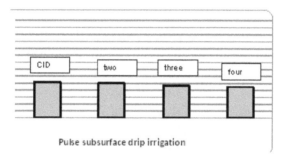

FIGURE 7.29 Effect of pulse subsurface drip irrigation on moisture content in root zone (MCRZ, Y-axis) under 50% of actual irrigation requirements.

in horizontal direction than vertical direction. These results are agreement with those obtained by El-Adi [21], Segal et al. [68], RO-DRIP® User Manual [65], Helen [40], Helmy [42], Shock et al. [70], Bouma et al. [10] Eric [28], and Zin El-Abedin [85]. Not only soil moisture content in the root zone was increased by increasing number of pulses but also pulse technique made enhancement in soil moisture distribution inside root zone and increased the $WSV_{\geq 100\%FC}$. The best conditions were determined according to the highest values of $WSV_{\geq 100\%FC}$, where increasing in $WSV_{\geq 100\%FC}$ means increasing in volume of available water in root zone.

Both MCRZ and $WSV_{\geq 100\%FC}$ were increased by increasing number of irrigation pulses, whereas MCRZ was increased from 10.04% under CDI to maximum value where it changed to 11.6% after applying pulse technique with 4 pulses at 100% of WRa under surface drip irrigation, recording an increase of 13.45%. Also, $WSV_{\geq 100\% FC}$ in root zone was increased from 8640 cm³ under CDI to a maximum value where it was 22320 cm³ after applying pulse technique with 4 pulses at 100% of WRa under surface drip irrigation, recording an increase of 61%.

TABLE 7.13 Moisture Content in Root Zone Under Subsurface Drip Irrigation at 50% of Actual Water Requirements

Soil depth	Distance from emitter (across lateral)			Distance from emitter (along lateral)			Moisture content in soil layer	Moisture content in root zone
	0	10	20	5	10	15		
	Continuous drip irrigation (CDI)							
	cm						%	
0	3.50	3.10	3.00	3.30	3.10	3.50		
15	12.00	9.70	5.00	11.00	9.70	5.00		
30	10.50	6.60	3.50	8.50	6.60	3.80		
45	10.00	4.50	3.80	7.20	4.50	3.50		
0–15	7.75	6.40	4.00	7.15	6.40	4.25	5.99	
15–30	11.25	8.15	4.25	9.75	8.15	4.40	7.66	
30–45	10.25	5.55	3.65	7.85	5.55	3.65	6.08	6.58
PSSDI, two pulses								
0	3.50	3.10	3.00	3.30	3.10	3.60		
15	12.00	9.40	5.00	11.50	9.70	5.20		
30	9.50	6.60	3.50	8.80	6.60	3.80		
45	9.40	4.50	3.80	5.50	4.50	3.50		
0–15	7.75	6.25	4.00	7.40	6.40	4.40	6.03	
15–30	10.75	8.00	4.25	10.15	8.15	4.50	7.63	
30–45	9.45	5.55	3.65	7.15	5.55	3.65	5.83	6.50
PSSDI, three pulses								
0	3.50	3.10	3.00	3.20	3.20	3.60		
15	12.00	7.40	6.00	10.30	9.70	5.20		
30	10.20	5.50	4.60	7.50	7.00	3.80		
45	9.50	4.50	3.30	5.00	4.70	3.50		
0–15	7.75	5.25	4.50	6.75	6.45	4.40	5.85	
15–30	11.10	6.45	5.30	8.90	8.35	4.50	7.43	
30–45	9.85	5.00	3.95	6.25	5.85	3.65	5.76	6.35
PSSDI, four pulses								
0	3.50	3.10	3.00	3.40	3.20	3.20		
15	12.00	8.00	6.30	9.00	9.50	5.20		
30	9.90	5.10	4.70	7.60	6.00	4.00		
45	9.50	4.50	3.30	5.00	4.70	3.50		
0–15	7.75	5.55	4.65	6.20	6.35	4.20	5.78	
15–30	10.95	6.55	5.50	8.30	7.75	4.60	7.28	
30–45	9.70	4.80	4.00	6.30	5.35	3.75	5.65	6.24

7.4.3 EFFECT OF PULSE DRIP IRRIGATION ON WATER APPLICATION EFFICIENCY (AE)

Application efficiency was calculated by dividing the volume of stored water in root zone by the volume of applied water. Tables 7.14 and Figures 7.30–7.31 show the relationship between pulse drip irrigation (PDI) and AE at 100%, 75% and 50% of WRa.

PSDI and AE: Table 7.14 and Figures 7.30 and 7.31 show that AE was increased by increasing number of pulses at 100% of WRa. AE was increased from 89% under CDI to 94% under pulse technique with 4 pulses,

TABLE 7.14 Application Efficiency Under Surface Drip Irrigation (PSDI) and Pulse Subsurface Drip Irrigation (PSSDI) at 100, 75%, and 50%% of Peak Actual Water Requirements (WRa) Under Continuous Drip Irrigation (CDI) and Number of Pulsing

Depth, cm	Application efficiency, %100%					
	At percentage of peak actual water requirements (WRa)					
	100	75	50	100	75	50
	Pulse surface drip irrigation, PSDI			Pulse subsurface drip irrigation, PSSDI		
CDI						
0–15	–	–	–	–	–	–
15–30	–	–	–	–	–	–
30–45	89	94	97	86	93	98.5
2 pulses						
0–15	–	–	–	–	–	–
15–30	–	–	–	–	–	–
30–45	92	92	96	88	94	98
3 pulses						
0–15	–	–	–	–	–	–
15–30	–	–	–	–	–	–
30–45	93.5	92	94	90.5	95.5	97.4
4 pulses						
0–15	–	–	–	–	–	–
15–30	–	–	–	–	–	–
30–45	94	97.5	91.5	93.4	96.2	97

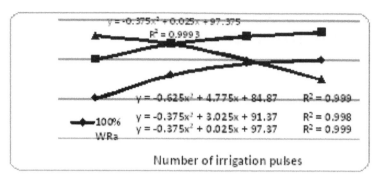

FIGURE 7.30   Effect of pulse surface drip irrigation on application efficiency (AE in %, Y-axis).

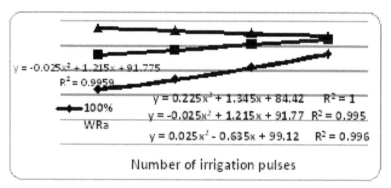

FIGURE 7.31   Effect of pulse subsurface drip irrigation on application efficiency (AE in %, Y-axis).

recording an increase of 5.3% of AE. AE was increased by increasing the number of pulses at 75% of WRa. AE was increased from 94% for CDI to 97.5% under pulse technique with 4 pulses, recording an increase of 3.6% of AE. AE was decreased by increasing number of pulses at 50% of WRa; and was decreased from 97% for CDI to 91.5 % under pulse technique with 4 pulses, recording a decrease of 6%.

PSSDI and AE: Table 7.14 and Figures 7.30 and 7.31 show that AE was increased by increasing number of pulses at 100% of WRa. AE was increased from 86% for CDI to 93.4% under pulse technique with 4 pulses, recording an increase of 7.9% of AE. AE was increased by increasing the number of pulses at 75% of WRa. AE was increased from 93% for CDI to 96.2% under pulse technique with 4 pulses, recording an increase of 3.3% of AE. AE was decreased by increasing number of pulses at 50% of WRa.

AE was decreased from 98.5% for CDI to 97% under pulse technique with 4 pulses, recording a decrease of 1.55%.

From the data in this section, we can conclude that AE was increased by increasing the number of irrigation pulses at 100% and 75% of WRa under surface and subsurface drip irrigation systems, due to decreasing of water movement downward by increasing of initial moisture content that was increased by increasing number of irrigation pulses. Hence, pulse technique can increase the water movement in horizontal direction than vertical direction. This action was also increased in wetted soil volume inside root zone thus meaning an increase in water volume which was stored in root zone. These results are in agreement with those reported by Scott [66] and Oron [61].

AE was decreased by increasing the number of irrigation pulses at 50% of WRa under surface and subsurface drip irrigation systems. This may be due to the ability of pulse technique to increase the wetted area of small amount due to applied water in addition increasing the time-off, hence the increasing the evaporation rate.

AE was decreased by increasing the amount of applied water. This is due to decrease in volume of stored water in root zone compared with increasing the volume of applied water. The values of AE under surface drip irrigation are higher than values of AE under subsurface drip irrigation, due to escaping the water out of root zone by deep percolation.

AE increased from 89% under CDI to a maximum value where it changed to 94% after applying pulse technique with 4 pulses at 100% of WRa under surface drip irrigation, recording an increase of 5.3%. On the other hand, maximum values of AE under deficit irrigation were 97.5% for 75% of WRa with 4 pulses under surface drip irrigation and 98.5% for 50% of WRa with CDI under subsurface drip irrigation.

7.4.4   EFFECTS OF PULSE DRIP IRRIGATION ON CLOGGING RATIO OF EMITTERS

Clogging ratio of emitters (CRE) is one of the important parameters for the evaluation of pulse drip irrigation performance. Clogging ratio of emitters was estimated from discharge of used emitters (after two years) and discharge of new emitters. Tables 7.15 and 7.16 and Figures 7.32 and 7.33 show the relationship between PDI and CRE under 100%, 75% and 50% of WRa.

TABLE 7.15 Effects of Number of Irrigation Pulses Under Surface Drip Irrigation (PSDI) on Clogging Ratio of Emitters at 100%, 75% and 50% of Actual Water Requirements

	Number of pulses in PSDI			
	4	3	2	Continuous
Clogging ratio of emitters, CRE				
100% WRa	5.50	6.60	7.93	9.85
75% WRa	5.45	6.52	7.86	9.81
50% WRa	5.38	6.49	7.85	9.79
L.S.D. = 0.12	ns	ns	ns	ns
Mean	5.44	6.54	7.88	9.83

TABLE 7.16 Effects of Number of Irrigation Pulses Under Subsurface Drip Irrigation (PSSDI) on Clogging Ratio of Emitters at 100%, 75% and 50% of Actual Water Requirements

	Number of pulses in PSSDI			
	4	3	2	Continuous
Clogging ratio of emitters, CRE				
100% WRa	7.00	8.11	9.40	10. 85
75% WRa	6.93	8.05	9.30	10. 78
50% WRa	6.89	7.99	9.29	10.73
L.S.D. = 0.12	ns	ns	ns	ns
Mean	6.94	8.05	9.33	10.79

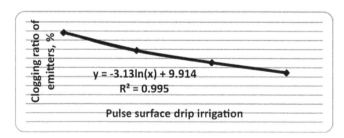

$$y = -3.13\ln(x) + 9.914$$
$$R^2 = 0.995$$

Pulse surface drip irrigation

FIGURE 7.32 Effects of number of irrigation pulses under surface drip irrigation on clogging ratio of emitters.

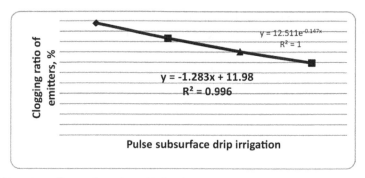

FIGURE 7.33 Effects of number of irrigation pulses under subsurface drip irrigation on clogging ratio of emitters.

PSDI and CRE at 100%, 75% and 50% of WRa (Table 7.15 and Figure 7.32): CRE was decreased by increasing the number of pulses at 100%, 75%, and 50% of WRa. CRE was decreased from 9.85% for CDI to 5.50% under pulse technique with 4 pulses, recording a decrease of 44%. At 75% of WRA, CRE was decreased from 9.81% for CDI to 5.45% under pulse technique with 4 pulses, recording a decrease of 44%. At 50% of WRa, CRE was decreased from 9.79% for CDI to 5.38% under pulse technique on 4 pulses, recording a decrease of 45%. There were no significant differences among values of CRE and amount of applied water. This means that no relationship existed between amount of water and CRE. It was possible to take the average of these values to form a general relationship (Logarithmic model) between CRE and NIP (number of irrigation pulses, 1 to 23) in for sandy soils under surface drip irrigation system:

$$CRE = -[3.13] \log_e (NIP) + 9.914 \tag{21}$$

The limitations of Eq. (21) are:

- for surface drip irrigation system;
- for sandy soils;
- range of irrigation pulses number (from 1 to 23 pulses);
- under organic agriculture;
- the kind of emitter is built in with discharge 2.1 L/h at 1 bar;
- time-off between pulses 30 minutes.

PSSDI and CRE at 100%, 75% and 50% of WRa (Table 7.16 and Figure 7.33): CRE was decreased by increasing the number of pulses at

100% of WRa. CRE was decreased from 10.85% for CDI to 7% under pulse technique with 4 pulses, recording an increase of 35%. The CRE was decreased by increasing the number of pulses at 75% and 50% of WRa. At 75%: CRE was decreased from 10.78% for CDI to 6.93% under pulse technique with 4 pulses, recording a decrease of 36%. At 50% of WRa: CRE was decreased from 10.73% for CDI to 6.89% under pulse technique with 4 pulses, recording a decrease of 36%. Under PSSDI, there were no significant differences among the values of CRE and amount of applied water. This means that under SSDI no relationship existed between amount of water and CRE. Therefore, it is possible to form a general relationship (linear model) between CRE and NIP (1 to 9 pulses) in sandy soils under subsurface drip irrigation system:

$$CRE = - [1.283] \, NIP + 11.98 \qquad (22)$$

The limitations of Eq. (22) are:

- for subsurface drip irrigation system for sandy soils;
- range of irrigation pulses number (from 1 to 9 pulse);
- under organic agriculture;
- the kind of emitter is built in with discharge 2.1 lph at 1 bar;
- time-off between pulses 30 minutes.
- maximum operating time per one pulse is 75 minutes.

Maximum value of CRE was 10.85% under 100% of WRa with CDI under SSDI and minimum value of CRE was 5.38% under 50% of WRa with 4 pulses under SDI. In general, CRE was decreased by increasing the number of irrigation pulses at 100%, 75% and 50% of WRa, under surface and subsurface drip irrigation. This may be due to the fact that pulse technique (cycle on-off) was able to create turbulence in the flow channel and this turbulence prevented particles in suspension from accumulation in flow channel and emitters. These results are agreement with those obtained by Solomon [73], Kolganov and Nosenko [51], Yardeni [83], Al-Amoud and Saeed [4] and Jackson [44].

CRE was increased by increasing the amount of applied water. This may be due to two reasons: Increasing amount of applied water meant increasing in number of operating hours hence increasing in probability of clogging; and increasing amount of applied water meant increasing in relative saturation of soil around emitter hence easily pull particles suspended in soil water during cycle on-off.

The values of CRE under pulse surface drip irrigation were lower than values of CRE under pulse subsurface drip irrigation, because pulse technique makes vacuum in the network and this vacuum with subsurface drip system will pull more particles around the emitters.

Using statistical analysis for values of CRE indicated that there are significant differences between pulse drip irrigation and continuous drip irrigation (L.S.D. at 5% level was 0.12).

7.4.5 EFFECT OF PULSE DRIP IRRIGATION ON EMISSION UNIFORMITY

In drip irrigation, water is carried in a pipe network to the point where it infiltrates into the soil. Therefore, the uniformity of application depends on the uniformity of emitter discharges throughout the system. Non-uniform discharge is caused by differences due to friction loss and elevation, variations between emitters due to manufacturing tolerances and clogging, pressure variations. Emission uniformity of drip irrigation system is a measure of the uniformity of emissions from all the emission points in the field.

Emission uniformity was calculated by dividing average rate of emitter discharge readings of the lowest one-fourth of the field data by average discharge rate of the emitters in the field. Tables 7.17 and Figures 7.34 and 7.35 show the relationship between pulse drip irrigation and emission uniformity under 100%, 75% and 50% of actual water requirements (WRa).

PSDI and EU under 100%, 75% and 50% of WRa: EU was increased by increasing number of pulses at 100%, 75% and 50% of Wra. At 100% of Wra, EU increased from 84.38% for continuous drip irrigation to 89.81% under pulse technique with 4 pulses, recording an increase of 6.4%. At 75% from Wra, EU increased from 84.95% for CDI to 90.26% under pulse technique with 4 pulses, recording an increase of 6.3 %. At 50% from Wra, EU increased from 85.02% for CDI to 90.48% under pulse technique with 4 pulses, recording an increase of 6.4%.

PSSDI and EU under 100%, 75% and 50% of WRa: EU was increased by increasing the number of pulses at 100%, 75% and 50% of WRa. At 100% of WRa, EU increased from 81.37% for CDI to 84.88%

TABLE 7.17 Effects of Pulse Surface Drip Irrigation (PSDI) and Subsurface Drip Irrigation (PSSDI) on Emission Uniformity

WR_{DI} (m³)	Number of Pulses	q_{min} L/h	q_a (L/h)	Emission uniformity (%)
Surface drip irrigation, PSDI				
3476	CDI	1.62	1.92	84.38
	2 Pulses	1.74	2.02	86.14
	3 Pulses	1.88	2.11	89.10
	4 Pulses	1.94	2.16	89.81
2707	CDI	1.75	2.06	84.95
	2 Pulses	1.67	1.93	86.53
	3 Pulses	1.91	2.14	89.25
	4 Pulses	1.76	1.95	90.26
1938	CDI	1.76	2.07	85.02
	2 Pulses	1.85	2.11	87.67
	3 Pulses	1.81	2.02	89.60
	4 Pulses	1.90	2.10	90.48
Subsurface drip irrigation, PSSDI				
3476	CDI	1.66	2.04	81.37
	2 Pulses	1.52	1.83	83.06
	3 Pulses	1.54	1.84	83.70
	4 Pulses	1.74	2.05	84.88
2707	CDI	1.69	2.03	83.25
	2 Pulses	1.76	2.11	83.41
	3 Pulses	1.72	2.04	84.31
	4 Pulses	1.89	2.18	86.70
1938	CDI	1.81	2.17	83.41
	2 Pulses	1.79	2.12	84.43
	3 Pulses	1.78	2.09	85.17
	4 Pulses	1.88	2.15	87.44

Note: Average of the two growing seasons; q_{min} = Minimum emitter discharge of low quarter, q_a = The average discharge rate of all emitters, and WR_{DI} = water regime under deficit irrigation.

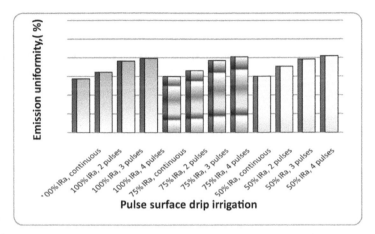

FIGURE 7.34 Effects of pulse surface drip irrigation on emission uniformity: PSDI.

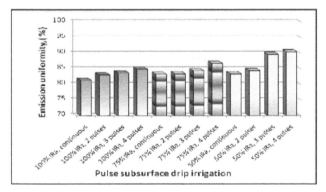

FIGURE 7.35 Effects of pulse subsurface drip irrigation on emission uniformity: PSSDI.

under pulse technique with 4 pulses, recording an increase of 4.3%. At 75% of WRa, EU increased from 83.25% for CDI to 86.70% under pulse technique with 4 pulses, recording an increase of 4.1%. At 50% of WRA, EU increased from 83.41% for CDI to 87.44% under pulse technique with 4 pulses, recording an increase of 4.8%. Maximum value of EU was 90.48% at 50% of WRa with 4 pulses under PSDI and minimum value of EU was 81.37% at 100% from WRa with CDI under PSSDI.

Based on results in this section, we can conclude that EU was increased by increasing the number of irrigation pulses (at 100%, 75% and 50% of WRa) under surface and subsurface drip irrigation systems. This was due to the positive effect of pulse drip irrigation on reducing CRE as shown in Figures 7.36 and 7.37. These results are in agreement with those obtained by Bader et al. [5]. EU increased from 85.02% under CDI to maximum value, where it changed to 90.48 % after applying pulse technique with 4 pulses at 50% of WRa under SDI, recording an increase 6.4%.

7.4.6 EFFECTS OF PULSE DRIP IRRIGATION ON POTATO YIELD

The relationship between pulse drip irrigation and potato yield (tons/feddan) under 100%, 75% and 50% of WRa are shown in Figures 7.38 and 7.39. Under PSDI (Figure 7.38), yield was increased by increasing the number of pulses at 100% and 75% of WRa. At 100% WRa, yield increased from 4.35 (ton/fed.) for CDI to 6.50 (ton/fed.) under pulse technique with

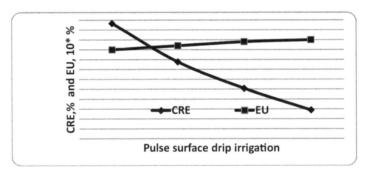

FIGURE 7.36 Effects of pulse surface drip irrigation on emission uniformity and clogging ratio of emitters: PSDI.

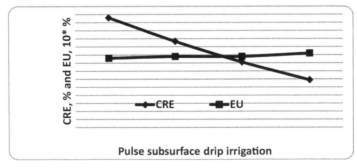

FIGURE 7.37 Effects of pulse subsurface drip irrigation on emission uniformity and clogging ratio of emitters: PSSDI.

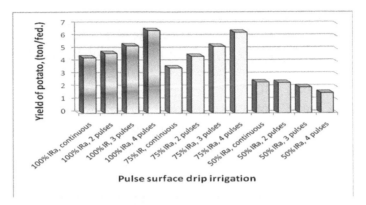

FIGURE 7.38   Effects of pulse surface drip irrigation on potato yield.

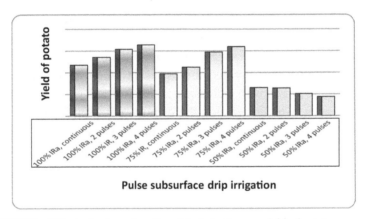

FIGURE 7.39   Effect of pulse subsurface drip irrigation on yield of potato.

4 pulses, recording an increase of 49%. At 75% of WRa, yield increased from 3.54 (ton/fed.) for CDI to 6.35 (ton/fed.) under pulse technique with 4 pulses, recording an increase of 79%. At 50% of WRa, yield was decreased by increasing the number of pulses. It decreased from 2.41 (ton/fed.) for CDI to 1.56 (ton/fed.) under pulse technique with 4 pulses, recording a decrease of 35.3%.

Figure 7.39 for PSSDI indicates that yield was increased by increasing the number of pulses at 100% and 75% of WRa. Yield increased from 4.70 (ton/fed.) for CDI to 6.57 (ton/fed.) under pulse technique with 4 pulses, recording an increase of 40%, at 100% of WRa. At 75% of WRa, yield increased from 3.89 (ton/fed.) for CDI to 6.39 (ton/fed.) under pulse technique with 4 pulses, recording an increase of 64%. At 50% of WRa, yield decreased from 2.60 (ton/fed.) for CDI to 1.75 (ton/fed.) under pulse technique with 4 pulses, recording a decrease of 32.7%.

Maximum value was 6.57 (ton/fed.) at 100% of WRa with 4 pulses under PSSDI and minimum value was 1.56 (ton/fed.) at 50% from WRa with 4 pulses under PSDI.

At 100% and 75% of WRa, the increase in yield was due to increase in the available nutrients in the root zone. These nutrients are more available for plant by increasing wetted soil volume (more than or equal 100% of field capacity) and moisture content in root zone. Previous data indicated the positive role of pulse drip irrigation with increasing AE. These results are in agreement with those obtained by Zin El-Abedin [85], Feng-Xin [30], Segal et al. [68], Beeson [8] and Nosenko, et al. [60]. Values of AE at 50% of WRa under surface and subsurface drip irrigation systems were very high but the depth of water stored was not sufficient for growing of potato plant and made high stress for plant and the harm effect of pulse drip irrigation with small amount of water and increasing in time-off. Small amount of water with more pulses with increasing in time-off will concentrate the salts around the plant, which increases the osmotic potential hence increase in probability of plasmolysis (Loss of water through osmosis is accompanied by shrinkage of protoplasm away from the cell wall). Hence, yield of potato was decreased especially under surface drip irrigation.

The statistical analysis for values of yield indicated that there are significant differences between yield values in pulse drip irrigation and continuous drip irrigation (L.S.D. at 5% level was 0.19). Yield was increased by increasing number of irrigation pulses at 100% and 75% of WRa). Yield increased from 4.70 (ton/fed.) under CDI to maximum value, where it changed to 6.57 (ton/fed.) after applying pulse technique with 4 pulses at 100% of WRa under subsurface drip irrigation, recording an decrease 40%. There were no significant differences between 6.50 (ton/fed.) at 100% of WRa with 4 pulses under SDI) and 6.39 (ton/fed.) at 75% of WRa with 4 pulses under SSDI, respectively.

7.4.7 EFFECTS OF PULSE DRIP IRRIGATION ON WATER USE EFFICIENCY

Water use efficiency (WUE) is an indicator of effective use of irrigation water unit for increasing crop yield. The WUE_{potato} was calculated by dividing total yield by total applied irrigation water. WUE_{potato} was evaluated for drip irrigation systems (surface and subsurface), water regimes under deficit irrigation (100%, 75% and 50% of WRa) and pulse irrigation (once per day = CDI, 2 pulses per day, 3 pulses per day and 4 pulses per day), as shown in Table 7.18.

Pulse surface drip irrigation (PSDI) and WUE_{potato} under 100%, 75% and 50% of WRa (Table 7.18): The WUE_{potato} was increased by increasing the number of pulses at 100% and 75% of WRa. At 75% of WRA, WUE_{potato} increased from 1.25 kg/m³ for continuous drip irrigation to 1.87 kg/m³ under pulse technique with 4 pulses, recording an increase of 49.6%. At 75% of WRa, WUE_{potato} increased from 1.31 kg/m³ for CDI to 2.35% under pulse technique with 4 pulses, recording an increase of 79.4%. WUE_{potato} was decreased by increasing number of pulses at 50% of WRa. WUE_{potato} decreased from 1.24 kg/m³ for CDI to 0.80 (kg/m³) under pulse technique with 4 pulses, recording a decrease of 35.5%.

Pulse subsurface drip irrigation and WUE_{potato} under 100%, 75% and 50% of WRa: Table 7.18 shows that WUE_{potato} was increased by increasing the number of pulses at 100% and 75% of WRa. WUE_{potato} increased from 1.35 (kg/m³ for CDI to 1.89 kg/m³ under pulse technique with 4 pulses, recording an increase of 40% at 100% of WRa. WUE_{potato} increased from 1.44 kg/m³ for CDI to 2.36 kg/m³ under pulse technique with 4 pulses, recording an increase of 63.9% at 75% of WRa. However, WUE_{potato} was decreased by increasing the number of pulses at 50% of WRa. WUE_{potato} decreased from 1.34 kg/m³ for CDI to 0.90 kg/m³ under pulse technique with 4 pulses, recording a decrease of 32.8%.

Maximum value of WUE_{potato} was 2.36 kg/m³ at 75% of WRa with 4 pulses under SSDI and minimum value of WUE_{potato} was 0.80 kg/m³ at 50% of WRa with 4 pulses under SDI. In general, there are significant differences between 2.36 kg/m³ at 75% of WRa with 4 Pulses under SSDI and 1.89 kg/m³ at 100% of WRa with 4 Pulses under SSDI. This may be due to the reduced amount of water applied that did not change the soil moisture significantly. Yields for both treatments were similar. These results are in agreement with those by Kenig et al. [47] and Segal et al. [68]. Statistical analysis for values of WUE_{potato} indicated that there are significant differences between pulse drip irrigation and continuous drip irrigation (L.S.D. at 5% level was 0.08).

WUE_{potato} was increased by increasing number of irrigation pulses especially at 100% and 75% from WRa. WUE_{potato} increased from

TABLE 7.18 Effects of Pulse Surface Drip Irrigation (PSDI) and Subsurface Drip Irrigation (PSSDI) on Water Use Efficiency of Potato: Average of Two Growing Seasons

WRDI (m3)	Number of Pulses	Y., (kg/fed.)	TAW (m³/fed./season)	WUE (kg/m³)
Surface drip irrigation, PSDI				
3476	CDI	4350	3476	1.25
	2 Pulse	4690	3476	1.35
	3 Pulse	5290	3476	1.52
	4 Pulse	6500	3476	1.87
2707	CDI	3540	2707	1.31
	2 Pulse	4450	2707	1.64
	3 Pulse	5240	2707	1.94
	4 Pulse	6350	2707	2.35
1938	CDI	2410	1938	1.24
	2 Pulse	2360	1938	1.22
	3 Pulse	2010	1938	1.04
	4 Pulse	1560	1938	0.80
L.S.D. at 5% level				0.08
Subsurface drip irrigation, PSSDI				
3476	CDI	4700	3476	1.35
	2 Pulse	5420	3476	1.56
	3 Pulse	6160	3476	1.77
	4 Pulse	6570	3476	1.89
2707	DI	3890	2707	1.44
	2 Pulse	4500	2707	1.66
	3 Pulse	5900	2707	2.18
	4 Pulse	6390	2707	2.36
1938	CDI	2600	1938	1.34
	2 Pulse	2560	1938	1.32
	3 Pulse	2040	1938	1.05
	4 Pulse	1750	1938	0.90
L.S.D. at 5% level				0.08

Y = Yield of potato; WUE = Water Use Efficiency; SDI = Surface Drip Irrigation; WR_{DI} = Water regime under deficit irrigation; IS = Irrigation Systems; TAW = Total applied water.

1.44 (kg/m³ under CDI to maximum value, where it changed to 2.36 kg/m³ after applying pulse technique with 4 pulses at 75% from WRa under SSDI, recording an increase 63.9%. This means that one can save 25% of actual water requirements per season, which is equivalent to 769 m³ of irrigation water.

7.4.8   EFFECTS OF PULSE DRIP IRRIGATION ON ENERGY USE EFFICIENCY

Energy use efficiency (EUE) is an indicator of effective use of energy unit in crop production. EUE_{potato} was calculated by dividing total yield by total energy consumption per season. EUE_{potato} was evaluated for: drip irrigation systems (surface and subsurface), water regime under deficit irrigation (100%, 75% and 50% of WRa) and pulse irrigation (once per day = CDI, 2 pulses per day, 3 pulses per day and 4 pulses per day). Total consumption energy was estimated by estimation the power requirements and operating time per season as shown in Tables 7.19 and 7.20.

Pulse surface drip irrigation and EUEpotato under 100%, 75% and 50% of WRa (Tables 7.19 and 7.20): EUE_{potato} was increased by increasing the number of pulses at 100% and 75% of WRa. EUE_{potato} increased from 2.94 kg/kw.h for continuous drip irrigation to 4.39 kg/kw.h under pulse technique with 4 pulses, recording an increase of 49% at 100% of WRa. At 75% of WRa, EUE_{potato} increased from 3.07 kg/kw.h for CDI to 5.51 kg/kw.h under pulse technique with 4 pulses, recording an increase of 79.5%. However, at 50% of WRa, EUE_{potato} was decreased by increasing number of pulses. EUE_{potato} decreased from 2.93 kg/kw.h for CDI to 1.90 kg/kw.h under pulse technique with 4 pulses, recording a decrease of 35%.

Pulse subsurface drip irrigation and EUEpotato under 100%, 75% and 50% of WRa (Tables 7.19 and 7.20): EUE_{potato} was increased by increasing the number of pulses at 100% and 75% of WRa. EUE_{potato} increased from 3.18 kg/kw.h for CDI to 4.44 kg/kw.h under pulse technique with 4 pulses, recording an increase of 39.6% at 100% of WRa. At 75% of WRa, EUE_{potato} increased from 3.37 kg /kw.h for CDI to 5.54 kg/kw.h under pulse technique with 4 pulses, recording an increase of 64%. However, at 50% of WRa, EUE_{potato} was decreased by increasing number of pulses. EUE_{potato} decreased from 3.16 kg/kw.h for CDI to

TABLE 7.19   Estimation of Energy Consumption Per Season

No.	Items		Value
1	**Power requirement**	Discharge = 45 m³/h	17.88
2	**(kw)**	Total dynamic head = 105 m	
3		Pump efficiency = 80 %	
4		Water specific weight =9.81 kN/m³	
5		Electric engine efficiency = 90 %	
6	**Operating time**	Operating time for 3476 m³/season	82.8
7	**(hours/season)**	Operating time for 2707 m³/season	64.5
8		Operating time for 1938 m³/season	46
9	**Energy consumption for 3476 m³/season**		1480 kw.h
10	**Energy consumption for 2707 m³/season**		1153 kw.h
11	**Energy consumption for 1938 m³/season**		822 kw.h

2.13 kg/kw.h under pulse technique with 4 pulses, recording a decrease of 32.6%. Maximum value of EUE_{potato} was 5.54 kg/kw.h at 75% of WRa with 4 pulses under SSDI and minimum value of EUE_{potato} was 1.9 kg/kw.h at 50% ofWRa with 4 pulses under SDI.

In general, EUE_{potato} was increased by increasing number of pulses at 100% and 75% of WRa. This may be due to two reasons: Increasing the yield by increasing irrigation pulses; and operating time will increase with increasing amount of applied water. These results are in agreement with those obtained by Georgiev [33] and Georgiev and Conley [32]. The statistical analysis for values of EUE_{potato} indicated that there are significant differences between pulse drip irrigation and continuous drip irrigation at 100% and 75% of WRa (L.S.D. at 5% level was 0.19). EUE_{potato} was increased from 3.37 kg/kw.h under CDI to maximum value, where it changed to 5.54 kg/kw.h after applying pulse technique with 4 pulses at 75% of WRa under SSDI, recording a decrease of 64%. This means that one can save 25% of energy consumption per season, which is equivalent to 327 kw.h. There were no significant differences between maximum value of EUE_{potato} and 5.51 kg/kw.h at 75% of WRa with 4 pulses under SDI.

TABLE 7.20 Effects of Pulse Surface Drip Irrigation (PSDI) and Pulse Subsurface Drip Irrigation (PSSDI) on Energy Use Efficiency: Average of Two Growing Seasons

WR_{DI} (m^3)	Number of Pulses	EUE_{potato} (kg/kw.h)
Surface drip irrigation, PSDI		
3476	CDI	2.94
	2 Pulses	3.17
	3 Pulses	3.57
	4 Pulses	4.39
2707	CDI	3.07
	2 Pulses	3.86
	3 Pulses	4.54
	4 Pulses	5.51
1938	CDI	2.93
	2 Pulses	2.87
	3 Pulses	2.45
	4 Pulses	1.90
L.S.D. at 5% level		0.18
Subsurface drip irrigation, PSSDI		
3476	CDI	3.18
	2 Pulses	3.66
	3 Pulses	4.16
	4 Pulses	4.44
2707	CDI	3.37
	2 Pulses	3.90
	3 Pulses	5.12
	4 Pulses	5.54
1938	CDI	3.16
	2 Pulses	3.11
	3 Pulses	2.48
	4 Pulses	2.13
L.S.D. at 5% level		0.19

7.4.9 EFFECTS OF PULSE DRIP IRRIGATION ON FERTILIZER USE EFFICIENCY

Fertilizer use efficiency (FUE) is an indicator of effective use of fertilizer unit in crop production. Fertilizer use efficiency of potato was calculated by dividing total yield by total amount of fertilizer application (N, P and K). NUE_{potato}, PUE_{potato} and KUE_{potato} were studied for: drip irrigation systems (surface and subsurface), water regime under deficit irrigation (100%, 75% and 50% of WRa) and pulse irrigation (once per day = CDI, 2 pulses per day, 3 pulses per day and 4 pulses per day). Total amount of N, P and K was 225 kg, 210 kg and 222.5 kg, respectively. Values of NUE_{potato}, PUE_{potato} and KUE_{potato} were calculated as shown in Table 7.21.

Under PSDI and 100% of WRa, Table 7.21 indicates that NUE_{potato}, PUE_{potato} and KUE_{potato} were increased by increasing number of pulses. The NUE_{potato} was increased from 19.55 kg/kg-N under continuous drip irrigation but it changed to 29.21 kg/kg-N after applying the pulse technique with 4 pulses. PUE_{potato} was increased from 20.71 kg/kg-P under CDI but it changed to 30.95 kg/kg-P after applying the pulse technique with 4 pulses. KUE_{potato} was increased from 19.33 kg/kg-K under CDI but it changed to 28.89 kg/kg-K after applying the pulse technique with 4 pulses.

Under PSSDI and 100% WRa, Table 7.21 indicates that NUE_{potato}, PUE_{potato} and KUE_{potato} were increased by increasing number of pulses. NUE_{potato} increased from 21.12 kg/kg-N under continuous drip irrigation to 29.53 kg/kg-N after applying the pulse technique with 4 pulses. PUE_{potato} increased from 22.38 kg/kg-P under CDI to 31.29 kg/kg-P after applying the pulse technique with 4 pulses. KUE_{potato} increased from 20.89 kg/kg-K under CDI to 29.20 kg/kg-K after applying the pulse technique with 4 pulses.

Under PSDI and 75% of WRa: Table 7.21 indicates that NUE_{potato}, PUE_{potato} and KUE_{potato} were increased by increasing number of pulses. NUE_{potato} increased from 15.91 kg/kg-N under continuous drip irrigation to 28.54 kg/kg-N after applying the pulse technique with 4 pulses. PUE_{potato} increased from 16.86 kg/kg-P under CDI to 30.24 kg/kg-P after applying the pulse technique with 4 pulses. KUE_{potato} increased from 15.73 kg/kg-K under CDI to 28.22 kg/kg-K after applying the pulse technique with 4 pulses.

TABLE 7.21 Effects of Irrigation Systems and Number of Pulses on Fertilizers Use Efficiency of Potato: Average of the Two Growing Seasons

I.S	WR$_{DI}$ (m^3)	Number of Pulses	NUE, (kg/kg-N)	PUE, (kg/kg-P)	KUE, (kg/kg-K)
PSDI	3476 = 100%	CDI	19.55	20.71	19.33
		2 Pulses	21.08	22.33	20.84
		3 Pulses	23.78	25.19	23.51
		4 Pulses	29.21	30.95	28.89
	2707= 75%	CDI	15.91	16.86	15.73
		2 Pulses	20.00	21.19	19.78
		3 Pulses	23.55	24.95	23.29
		4 Pulses	28.54	30.24	28.22
	1938 = 50%	CDI	10.71	11.48	10.83
		2 Pulses	10.61	11.24	10.49
		3 Pulses	9.03	9.57	8.93
		4 Pulses	7.01	7.43	6.93
PSSDI	3476 = 100%	CDI	21.12	22.38	20.89
		2 Pulses	24.36	25.81	24.09
		3 Pulses	27.69	29.33	27.38
		4 Pulses	29.53	31.29	29.20
	2707 = 75%	CDI	17.48	18.52	17.29
		2 Pulses	20.22	21.43	20.00
		3 Pulses	26.52	28.10	26.22
		4 Pulses	28.72	30.43	28.40
	1938 = 50%	CDI	11.56	12.38	11.69
		2 Pulses	11.51	12.19	11.38
		3 Pulses	9.17	9.71	9.07
		4 Pulses	7.87	8.33	7.78
L.S.D. at 5% level			0.84	0.90	0.85

NUE = Nitrogen Use Efficiency, PUE = Phosphorus Use Efficiency, KUE = Potassium Use Efficiency, PSDI = Surface drip irrigation, PSSDI = Subsurface drip irrigation, WR$_{DI}$ = Water regime under deficit irrigation, IS= Irrigation systems.

Under PSSDI and 75% of WRa: Table 7.21 shows that NUE_{potato}, PUE_{potato} and KUE_{potato} were increased by increasing number of pulses. The NUE_{potato} increased from 17.48 kg/kg-N under continuous drip irrigation to 28.72 kg/kg-N after applying the pulse technique with 4 pulses. PUE_{potato} increased from 18.52 (kg/kg-P) under CDI to 30.43 kg/kg-P after applying the pulse technique with 4 pulses. KUE_{potato} increased from 17.29 kg/kg-K under CDI to 28.40 kg/kg-K after applying the pulse technique with 4 pulses.

Under PSDI and 50% of WRa: Table 7.21 indicates that NUE_{potato}, PUE_{potato} and KUE_{potato} were decreased by increasing number of pulses. NUE_{potato} decreased from 10.71 kg/kgN under continuous drip irrigation to 7.01 kg/kg-N after applying the pulse technique with 4 pulses. PUE_{potato} decreased from 11.48 kg/kg-P under CDI to 7.43 kg/kg-P after applying the pulse technique with 4 pulses. KUE_{potato} decreased from 10.83 kg/kg-K under CDI to 6.93 kg/kg-K after applying the pulse technique with 4 pulses.

Under PSSDI and 50% of WRa: According to data in Table 7.21, NUE_{potato}, PUE_{potato} and KUE_{potato} were decreased by increasing number of pulses. NUE_{potato} decreased from 11.56 kg/kg-N under continuous drip irrigation to 7.87 kg/kg-N after applying the pulse technique with 4 pulses. PUE_{potato} decreased from 12.38 kg/kg-P under CDI to 8.33 kg/kg-P after applying the pulse technique with 4 pulses. KUE_{potato} decreased from 11.69 kg/kg-K under CDI to 7.78 kg/kg-K after applying the pulse technique with 4 pulses.

Maximum values of NUE_{potato}, PUE_{potato} and KUE_{potato} were 29.53 kg/kg-N, 31.29 kg/kg-P and 29.20 kg/kg-K, respectively at 100% of WRa with 4 pulses under SSDI and minimum values were 7.01 kg/kg-N, 7.43 kg/kg-P and 6.93 kg/kg-K at 50% of WRa with 4 pulses under SDI.

It can be concluded from the data on fertilizers use efficiency (nitrogen, phosphorus and potassium use efficiencies) in Table 7.21 that FUE was increased by increasing number of irrigation pulses at 100% and 75% from WRa under surface and subsurface drip irrigation. This was due to the positive effect of pulse drip irrigation on increasing the wetted soil volume in root zone hence increasing the ability of roots to absorb more fertilizers. Statistical analysis for values of NUE_{potato}, PUE_{potato} and KUE_{potato} indicated that there are significant differences among pulse drip irrigation and continuous drip irrigation at 100%, 75% and 50% of WRa. The L.S.D. at 5% level was 0.84 for NUE_{potato}, 0.90 for PUE_{potato} and 0.85 for KUE_{potato}. NUE_{potato}, PUE_{potato} and KUE_{potato} were increased from 21.12 kg/kg-N, 22.38 kg/kg-P and 20.89 kg/kg-K under CDI to maximum values, where these changed to 29.53 kg/kg-N, 31.29 kg/kg-P and 29.20 kg/kg-K, respectively after applying pulse technique with 4 pulses at 100% of WRa under SSDI.

7.4.10   EFFECTS OF PULSE DRIP IRRIGATION ON QUALITY OF POTATO TUBERS

7.4.10.1   Effect of Pulse Drip Irrigation on Tuber Dry Matter

Table 7.22 shows the relationship between tuber fry matter (TDM) and pulse drip irrigation under 100%, 75% and 50% of actual water requirements (WRa), for two irrigation systems.

PSDI and WRa: TDM was increased by increasing the number of pulses at 100% and 75% of WRa. TDM increased from 14.83 gm for continuous drip irrigation to 18.78 gm under pulse technique with 4 pulses, recording an increase of 26.63% at 100% of WRa. TDM increased from 13.98 gm for CDI to 17.15 gm under pulse technique with 4 pulses, recording an increase of 22.67% at 75% of WRa. However, at 50% of WRa, TDM decreased from 13.02 gm for CDI to 10.00 gm under pulse technique with 4 pulses, recording a decrease of 23.2%.

PSSDI and WRa: TDM was increased by increasing the number of pulses at 100% and 75% of WRa. TDM increased from 14.73 gm for CDI to 18.99 gm under pulse technique with 4 pulses, recording an increase of 28.92% at 100% of WRa. TDM increased from 14.97 gm for CDI to 18.82 gm under pulse technique with 4 pulses recording an increase of 25.71% at 75% of WRa. However, at 50% of WRa, TDM decreased from 13.52 gm for CDI to 10.57 gm under pulse technique with 4 pulses recording a decrease of 21.8%.

Maximum value of TDM was 18.99 gm at 100% of WRa with 4 Pulses under PSSDI and minimum value of TDM was 10 gm at 50% of WRa with 4 pulses under PSDI.

7.4.10.2   Effect of Pulse Drip Irrigation on Tuber Specific Density

Table 7.22 shows the relationship between pulse drip irrigation and tuber specific density (TSD) at 100%, 75% and 50% of actual water requirements (WRa), for two irrigation systems.

TABLE 7.22 Effects of Irrigation Systems, Number of Pulses and Water Regimes (WR$_{DI}$: 100%, 75% and 50%) on Quality Characteristics of Potato Tubers

I.S.	WR$_{DI}$ (m³)	Number of Pulses	Tuber characteristics		
			Dry matter, (%)	Specific density, (gm/cm³)	Total carbohydrates, (%)
PSDI	3476	CDI	14.83	1.0545	39.71
		2 Pulses	16.14	1.0607	42.93
		3 Pulses	18.21	1.0705	44.46
		4 Pulses	18.78	1.0732	45.07
	2707	CDI	13.98	1.0505	38.79
		2 Pulses	16.20	1.0610	41.50
		3 Pulses	16.82	1.0639	42.14
		4 Pulses	17.15	1.0655	42.66
	1938	CDI	13.02	1.0459	39.00
		2 Pulses	12.11	1.0416	38.19
		3 Pulses	11.44	1.0384	38.11
		4 Pulses	10.00	1.0316	37.55
PSSDI	3476	CDI	14.73	1.0540	40.52
		2 Pulses	16.46	1.0622	43.84
		3 Pulses	18.68	1.0727	44.55
		4 Pulses	18.99	1.0742	45.58
	2707	CDI	14.97	1.0551	41.29
		2 Pulses	16.41	1.0620	41.60
		3 Pulses	17.73	1.0682	42.53
		4 Pulses	18.82	1.0734	43.14
	1938	CDI	13.52	1.0483	37.58
		2 Pulses	12.41	1.0430	37.56
		3 Pulses	12.00	1.0411	34.57
		4 Pulses	10.57	1.0343	34.00
L.S.D. at 5% level			0.91	0.0042	0.84

PSDI = Surface drip irrigation, PSSDI = Subsurface drip irrigation, WR$_{DI}$ = Water regime under deficit irrigation, IS = Irrigation systems.

PSDI and WRa: According to data for 100%, 75% and 50% of WRa in Table 7.22, it can be observed that TSD was increased by increasing the number of pulses at 100% and 75% of WRa. At 100% of WRa, the TSD increased from 1.0545 gm/cm³ for continuous drip irrigation to 1.0732 gm/cm³ under pulse technique with 4 pulses recording an increase of 1.77%. At 75% of WRa, TSD increased from 1.0505 gm/cm³ for CDI to 1.0655 gm/cm³ under pulse technique with 4 pulses recording an increase of 1.43%. However, at 50% of WRa, TSD was decreased by increasing the number of pulses. TSD decreased from 1.0459 gm/cm³ for CDI to 1.0316 gm/cm³ under pulse technique with 4 pulses recording a decrease of 1.37%.

PSSDI and WRa: According to data for 100%, 75% and 50% of WRa in Table 7.22, it can be observed that TSD was increased by increasing the number of pulses at 100% and 75% of WRa. TSD increased from 1.0540 gm/cm³ for CDI to 1.0742 gm/ cm³ under pulse technique with 4 pulses recording an increase of 1.92% at 100% of WRa. At 75% of WRa, TSD increased from 1.0551 gm/cm³ for CDI to 1.0734 gm/cm³ under pulse technique with 4 pulses recording an increase of 1.73%. However, at 50% of WRa, TSD was decreased by increasing number of pulses. TSD decreased from 1.0483 gm/cm³ for CDI to 1.0343 gm/cm³ under pulse technique with 4 pulses recording an decrease of 1.34%.

Maximum value of TSD was 1.0742 gm/cm³ at 100% of WRa with 4 pulses under PSSDI and minimum value of TSD was 1.0316 gm/cm³ at 50% of WRa with 4 pulses under PSDI.

7.4.10.3 Effect of Pulse Drip Irrigation on Tuber Total Carbohydrates

Table 7.22 shows the relationship between pulse drip irrigation and tuber total carbohydrates TTC) at 100%, 75% and 50% of actual water requirements (WRa).

PSDI and WRa (Table 7.22): TTC was increased by increasing number of pulses at 100% and 75% of WRa. At 100% WRa, TTC increased from 39.71% for continuous drip irrigation to 45.07% under pulse technique with 4 pulses recording an increase of 13.49%. At 75% of WRa, TTC increased from 38.79% for CDI to 42.66% under pulse technique with 4 pulses recording an increase of 9.98%. However, at 50% of WRa,

TTC was decreased by increasing number of pulses at 50% of WRa. TTC decreased from 39% for CDI to 37.55% under pulse technique with 4 pulses recording a decrease of 3.72%.

PSSDI and WRa (Table 7.22): TTC was increased by increasing number of pulses at 100% and 50% of WRa. At 100% of WRa, TTC increased from 40.52% for CDI to 45.58% under pulse technique on 4 pulses recording an increase of 12.49%. At 75% from WRa, TTC increased from 41.29% for CDI to 43.14% under pulse technique with 4 pulses recording an increase of 4.48%. However, at 50% of WRa, TTC was decreased by increasing number of pulses. TTC decreased from 37.58% for CDI to 34% under pulse technique with 4 pulses recording a decrease of 9.52%.

Maximum value of TTC was 45.58% at 100% of WRa with 4 pulses under PSSDI and minimum value of TTC was 34% at 50% of WRa with 4 pulses under PSDI.

It can be noticed in Table 7.22 that there is a effect of pulse drip irrigation on increasing the tuber dry matter, tuber specific density and tuber total carbohydrates compared with continuous drip irrigation especially at 100% and 75% of WRa. This may be due to increase of AE under pulse irrigation, which increased the ability of roots to absorb more fertilizer nutrients. Also, pulse irrigation increased the water balance inside root zone. In conclusion, pulse irrigation helped plants to grow faster and healthier than plants that were stressed. These results are in agreement with those obtained by Bravdo and Poebsting [11], Steyn, et al. [77, 78], Sean [67] and Beeson [8]. No clear trend was observed for effects of pulse drip irrigation on quality of potato at 50% of WRa. This may be due to small amount of applied water with increasing number of pulses that increased salt accumulation inside root zone. Statistical analysis for values of TDM, TSD and TTC indicated that there are significant differences between pulse drip irrigation and continuous drip irrigation especially at 100% and 75% of WRa. The L.S.D. at 5% level was 0.91 for TDM, 0.0042 for TSD and 0.84 for TTC.

TDM, TSD and TTC increased from 14.73%, 1.0540 gm/cm^3, 40.52% under CDI to maximum values, where these changed to 18.99%, 1.0742 gm/cm^3, 45.58%, respectively, after applying pulse technique with 4 pulses at 100% of WRa under PSSDI. This is in agreement with the data in Table 7.23 that indicates increasing dry weight of stems and leaves/plant

TABLE 7.23 Effects of Irrigation Systems and Water Regimes (WR$_{DI}$: 100%, 75%, 50%) on Dry Weight of Stems and Leaves/Plant: Average of the Two Growing Seasons

IS	WRDI (m³)	Number of Pulses	Dry weight of stems and leaves of plant, gm			Leaf area/plant, cm²		
			60 day after sowing	80 day after sowing	100 day after sowing	100 day after sowing	80 day after sowing	60 day after sowing
PSDI	3476	CDI	20.28	28.37	36.46	1630	1300	974
		2 Pulses	22.97	31.96	40.95	1755	1400	1049
		3 Pulses	23.09	32.12	41.15	1817	1450	1085
		4 Pulses	25.43	35.24	45.05	1850	1477	1106
	2707	CDI	19.92	27.90	35.87	1570	1255	940
		2 Pulses	21.62	30.16	38.70	1696	1357	1016
		3 Pulses	24.55	34.06	43.58	1830	1459	1093
		4 Pulses	24.27	33.69	43.11	1826	1461	1094
	1938	CDI	15.40	21.86	28.33	1539	1230	921
		2 Pulses	15.20	21.60	28.00	1508	1201	900
		3 Pulses	15.18	21.58	27.97	1256	1000	750
		4 Pulses	13.47	19.30	25.12	1231	980	734
PSSDI	3476	CDI	21.33	29.78	38.22	1668	1330	995
		2 Pulses	23.19	32.26	41.32	1754	1400	1050
		3 Pulses	25.06	34.75	44.44	1838	1471	1100
		4 Pulses	26.07	36.10	46.12	1875	1500	1124
	2707	CDI	20.99	29.32	37.65	1631	1305	977
		2 Pulses	24.27	33.69	43.11	1725	1380	1034
		3 Pulses	25.39	35.18	44.98	1841	1468	1100
		4 Pulses	24.80	34.40	44.00	1839	1470	1100
	1938	CDI	17.33	24.44	31.55	1560	1250	938
		2 Pulses	16.68	23.58	30.47	1509	1200	906
		3 Pulses	15.91	22.55	29.19	1457	1160	873
		4 Pulses	15.73	22.31	28.89	1378	1100	824

WR$_{DI}$ = Water regime under deficit irrigation, PSDI = Surface drip irrigation, PSSDI = Subsurface drip irrigation.

and leaves area/plant. The highest value of dry weight of stems and leaves /plant was 46.12 gm and the highest value of leaves area/plant was 1875 cm² under PSSDI + 100% of WRa + 4 pulses. The increase in leaves area/plant led to increase of photosynthesis process that increased yield and enhanced the quality of potato tubers.

7.4.11 ECONOMIC ANALYSIS

The prices were based on the prices by exporting companies in Egypt. The price of organic potato in 2007 was 5000 L.E./ton. Basic data for economic analysis is presented in Table 7.24. The data on total costs of inputs, total income of output and net income are presented in Table 7.25.

Table 7.25 shows the relationship between irrigation systems and net income (NI) under 100%, 75% and 50% of actual water requirements (WRa).

TABLE 7.24   Basic Data for Calculation of Net Income: Potato Production

Items		All treatments: Irrigation systems; Water regime treatments; Number of pulses		
List of inputs	Cost of Irrigation, LE/fed.	100%	75%	50%
		574	530	487
	Cost of Land preparation, LE/fed.	300		
	Cost of tuber seeds, LE/fed.	6900		
	Cost of compost, LE/fed	4944		
	Cost of Microbin, LE/fed	55		
	Cost of weed control, LE/fed.	160		
	Cost of pest control, LE/fed.	100		
	Cost of harvesting, LE/fed.	170		
	Rent (on season), LE/fed.	1920		
	Total cost for inputs, TCI, LE/fed.	15123	15079	15036
Output	Yield, ton/fed.	Y_n		
	Price, LE/ton.	5000		
	Total income for output, LE/fed.	$= Y_n \times 5000$		
Net income = list of outputs – list of inputs		$[Y_n \times 5000] - TCI$		

$Y_n = Y$ is yield, and n = number of treatment (from 1 to 24 treatments);

TCI = Total costs for inputs; The prices in 2006/2007: where US$ 1.00 = 5.5 L.E.

TABLE 7.25 Effects of Irrigation Systems, Number of Pulses and Water Regimes (WR$_{DI}$: 100%, 75%, 50%) on Total Costs, Total Income and Net Income for Potato Production: Average of Two Growing Seasons

I.S	WRDI (m³)	Number of Pulses	TCI LE/ fed. = A	Total output, LE/fed. = B	Net income, LE/fed. = B − A
PSDI	3476	CDI	15123	21750	6627
		2 Pulses	15123	23450	8327
		3 Pulses	15123	26450	11327
		4 Pulses	15123	32500	17377
	2707	CDI	15079	17700	2621
		2 Pulses	15079	22250	7171
		3 Pulses	15079	26200	11121
		4 Pulses	15079	31750	16671
	1938	CDI	15036	12050	−2986
		2 Pulses	15036	11800	−3236
		3 Pulses	15036	10050	−4986
		4 Pulses	15036	7800	−7236
PSSDI	3476	CDI	15123	23500	8377
		2 Pulses	15123	27100	11977
		3 Pulses	15123	30800	15677
		4 Pulses	15123	32850	17727
	2707	CDI	15079	19450	4371
		2 Pulses	15079	22500	7421
		3 Pulses	15079	29500	14421
		4 Pulses	15079	31950	16871
	1938	CDI	15036	13000	−2036
		2 Pulses	15036	12800	−2236
		3 Pulses	15036	10200	−4836
		4 Pulses	15036	8750	−6286
L.S.D. at 5% level					896

PSDI and WRa (Table 7.25 and Figure 7.40): Net income (NI) was increased by increasing number of pulses at 100% and 75% of WRa. At 100% WRa, NI increased from 6627 LE/fed. for continuous drip irrigation to 17377 LE/fed. under pulse technique with 4 pulses. At 75% of WRa, NI increased from 2621 LE/fed. for CDI to 16671 LE/fed. under pulse technique with 4 pulses. However, at 50% of WRa, value of NI was less than zero at 50%. NI decreased from −2986 LE/fed. for CDI to −7236 LE/fed. under pulse technique with 4 pulses.

PSSDI and WRa (Table 7.25 and Figure 7.41): NI was increased by increasing number of pulses at 100% and 50% of WRa. At 100% WRa, NI increased from 8377 LE/fed. for CDI to 17727 LE/fed. under pulse technique with 4 pulses. At 75% of WRa, NI increased from 4371 LE/fed. for CDI to 16871 LE/fed. under pulse technique with 4 pulses. However, at 50% of WRa, NI was less than zero. NI decreased from −2036 LE/fed. for CDI to −6286 LE/fed. under pulse technique with 4 pulses.

Maximum value of NI was 17727 LE/fed. at 100% of WRa with 4 pulses for PSSDI and minimum value of NI was −7236 LE/fed. at 50% of WRa with 4 pulses for SDI. The increase in NI was due to the increase in tuber yield. Statistical analysis for values of NI indicated that there are significant differences between pulse drip irrigation and continuous drip irrigation. The L.S.D. at 5% level was 896. NI increased from 8377 LE/fed. under CDI to maximum value, where it was 17727 LE/fed. after applying pulse technique with 4 pulses at 100% of WRa under PSSDI. There were no significant differences between maximum value of NI and 17377 LE/fed. and 16871 LE/fed. at 100% of WRa with 4 Pulses under PSDI and at 75% of WRa with pulses under PSSDI, respectively.

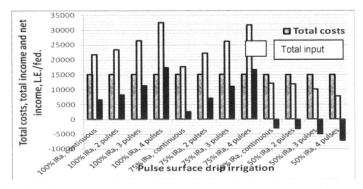

FIGURE 7.40   Effects of pulse surface drip irrigation on total costs, total income and net income (NI: solid bars).

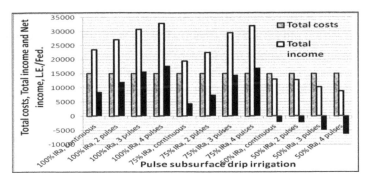

FIGURE 7.41 Effects of pulse subsurface drip irrigation on total costs, total income and net income (NI: solid bars).

7.5 CONCLUSIONS AND RECOMMENDATIONS

Moisture content in the root zone was increased by increasing number of irrigation pulses, where moisture content in the root zone was increased from 10.04% under continuous drip irrigation to maximum value where it changed to 11.6 % after applying pulse technique with 4 pulses at 100% of actual water requirements under surface drip irrigation, recording an increase of 13.45%. Wetted soil volume (more than or equal 100% of field capacity) in root zone was increased by increasing number of irrigation pulses where wetted soil volume in root zone increased from 8640 cm³ under continuous drip irrigation to maximum value of 22320 cm³ after applying pulse technique with 4 pulses at 100% of actual water requirements under surface drip irrigation, recording an increase of 61%. Therefore, it is recommended to use pulse drip irrigation to irrigate sandy soils.

Application efficiency was increased by increasing number of irrigation pulses at 100% and 75% of WRa under surface and subsurface drip irrigation systems. AE increased from 89% under CDI to maximum value of 94% after applying pulse technique with 4 pulses at 100% of WRa under surface drip irrigation, recording an increase of 5.3%. Maximum values of AE under deficit irrigation were 97.5% for 75% with 4 pulses under surface drip irrigation and 98.5% for 50% with CDI under subsurface drip irrigation.

Clogging ratio of emitters was decreased by increasing number of irrigation pulses. Clogging ratio of emitters decreased from 9.79% under continuous drip irrigation to minimum value of 5.38% after applying pulse technique with 4 pulses at 50% of actual water requirements under surface drip irrigation, recording a decrease 45%.

EU was increased by increasing number of irrigation pulses. EU increased from 85.02% under CDI to maximum value of 90.48 % after applying pulse technique with 4 pulses at 50% of WRa under SDI, recording an increase 6.4%.

Potato yield was increased by increasing number of irrigation pulses at 100% and 75% of actual water requirements. It increased from 4.70 ton/fed. under continuous drip irrigation to maximum value of 6.57 ton/fed. after applying pulse technique with 4 pulses at 100% of actual water requirements under subsurface drip irrigation, recording an increase 40%.

There were no significant differences between maximum yield and 6.50 ton/fed. and 6.39 ton/fed. at 100% of actual irrigation requirements with 4 pulses under surface drip irrigation and 75% of actual water requirements with 4 pulses under subsurface drip irrigation, respectively.

Avoid using pulse technique at 50% of actual water requirements. Because, increasing number of irrigation pulses with small amount of applied water, will increase salt accumulation inside the root zone, hence causing decrease in yield.

Water use efficiency of potato was increased by increasing number of irrigation pulses at 100% and 75% of actual water requirements. It increased from 1.44 kg/m^3 under continuous drip irrigation to maximum value of 2.36 kg/m^3 after applying pulse technique with 4 pulses at 75% of WRa under SSDI, recording an increase 63.9%. This implies a 25% water saving that is equivalent of 769 m^3 of water.

Energy use efficiency of potato was increased by increasing number of irrigation pulses especially at 100% and 75% of WRa. It increased from 3.37 kg/kw.h under CDI to maximum value of 5.54 kg/kw.h after applying pulse technique with 4 pulses at 75% of WRa under SSDI, recording a decrease of 64%. It amounted to 25% saving in energy consumption per season, which is equivalent to 327 kw.h. There were no significant differences between maximum value of energy use efficiency of potato and 5.51 kg/kw.h at 75% of WRa with 4 pulses under SDI.

Fertilizers use efficiency (nitrogen, phosphorus and potassium use efficiencies) were increased by increasing number of irrigation pulses at 100% and 75% of actual water requirements. FUE increased from 21.12 kg/kg-N,

22.38 kg/kg-P and 20.89 kg/kg-K under continuous drip irrigation to maximum values of 29.53 kg/kg-N, 31.29 kg/kg-P and 29.20 kg/kg-K, respectively after applying pulse technique with 4 pulses at 100% of actual water requirements under subsurface drip irrigation.

Tuber dry matter, tuber specific density and tuber total carbohydrates were increased by increasing number of irrigation pulses at 100% and 75% of actual water requirements. Tuber dry matter, tuber specific density and tuber total carbohydrates increased from 14.73%, 1.0540 gm/cm^3, 40.52% under continuous drip irrigation to maximum values of 18.99%, 1.0742 gm/cm^3, 45.58%, respectively after applying pulse technique with 4 pulses at 100% of actual water requirements under subsurface drip irrigation.

Net income was increased by increasing number of irrigation pulses at 100% and 75% of actual water requirements. Net income increased from 8377 LE/fed. under continuous drip irrigation to maximum value of 17727 LE/fed. after applying pulse technique with 4 pulses at 100% of actual water requirements under subsurface drip irrigation. From the results in this chapter, author recommended following:

a. To get maximum yield, quality of potato and net income, one must apply pulse technique with 4 pulses at 100% of actual water requirements under subsurface drip irrigation.
b. In the case of water and energy limitations, one has to apply pulse technique with 4 pulses at 75% of actual water requirements under subsurface drip irrigation.

7.6   SUMMARY

This study was conducted to study the performance of pulse drip irrigation under organic agriculture for potato crop in sandy soils for saving water and fertilizers, increasing yield of potato, increasing the energy use efficiency, improving potato quality, decreasing the costs and increasing income under Egyptian growing conditions. In the research study, author used two drip irrigation systems, three water application rates and three types for pulse irrigation. The experiment was carried out in a sandy soil during two summer growing seasons 2006 and 2007, in Abo-Ghaleb farm, Cairo- Alexandria Rood, 60 Km away from Cairo. The evaluation parameters were: (1) Soil moisture distribution, (2) Application efficiency (AE), (3) Clogging ratio of emitters (CRE), (4) Emission uniformity (EU), (5) Potato yield, (6) Water use efficiency of potato (WUE_{potato}), (7) Energy use efficiency (EUE), (8) Fertilizer use efficiency (FUE), (9) Quality characteristics of potato tubers, and (10) Net income (NI). The salient results are summarized as follows:

i. Moisture content in the root zone and wetted soil volume (more than or equal 100% of field capacity) in root zone were increased by increasing number of irrigation pulses.

ii. AE was increased from 89% under continuous drip irrigation to maximum value of 94% after applying pulse technique with 4 pulses at 100% of actual water requirements under surface drip irrigation, recording an increase of 5.3%.

iii. CRE was decreased from 9.79% under continuous drip irrigation to minimum value of 5.38% after applying pulse technique on 4 pulses at 50% of actual water requirements under surface drip irrigation, recording a decrease 45%.

iv. EU was increased from 85.02% under continuous drip irrigation to maximum value of 90.48% after applying pulse technique on 4 pulses at 50% of actual water requirements under surface drip irrigation, recording an increase 6.4 %.

v. Potato yield was increased from 4.70 ton/fed. under continuous drip irrigation to maximum valube of 6.57 ton/fed. after applying pulse technique with 4 pulses at 100% of actual water requirements under subsurface drip irrigation, recording an increase 40%.

vi. To get maximum yield, best quality characteristics of tubers and net income, we must apply pulse technique on 4 pulses at 100% of actual water requirements under subsurface drip irrigation.

vii. Under water and energy limitations, we have to apply pulse technique on 4 pulses at 75% of actual water requirements under subsurface drip irrigation.

ACKNOWLEDGMENTS

Author acknowledges support and advice of PhD Dissertation committee consisting of: Dr. Gomma Abdrabo Abd El-Rahman, Dr. Farthy Gad El-Ebabi and Dr. Mouhamad Talaat El-Saidi.

KEYWORDS

- actual water requirements
- clogging ratio of emitters
- deficit irrigation
- Egypt
- emitter clogging
- evapotranspiration
- feddan
- fertilizer use efficiency
- irrigation requirements
- L.E.
- net income
- nitrogen use efficiency
- phosphorous use efficiency
- potassium use efficiency
- potato
- pulse irrigation
- pulse micro irrigation
- sandy soil
- soil moisture
- soil moisture distribution
- subsurface drip irrigation
- surface drip irrigation
- water regime
- water use efficiency

REFERENCES

1. A. O. A. C. (1975). *Official Methods of Analysis 12th Ed.* Association of official Analysis Chemists. Washington – USA. 700 pp.
2. Abdel-Aal, El. I. (2000). Effect of some parameters of trickle irrigation system on pea production. *Misr. J. Ag. Eng.,* 17(1), 113–124.

3. Abou-Hussein, S. (2003). *Studies on Potato Production under Organic Conditions.* PhD, Horticulture Dept., Fac. of Agric., Ain Shams Univ., Egypt, 170 pp.

4. Al-Amoud, A. I., & Saeed, M. (1988). The effect of pulsed drip irrigation on water management. *Proceedings 4th International Micro irrigation Congress,* 4B-2, p. 10.

5. Bader, A. E., Bakeer, G. A., Abdelrahman, G. M., & Elsayed, T. S. (2001). Comparative study of some micro irrigation systems under semi-arid regions. *Misr. J. Ag. Eng.,* 18 (1), 151–168

6. Barber, S. A. (1976). Efficient fertilizer use. *Agronomic Research for Food. WIASA Special Publication No 26 by* Amer. Soc. Agron., pp. 13–29.

7. Bartok, J. W. (1989). Ebb and flow: new technology provides efficient irrigation alternatives. *Greenhouse Manager,* 8 (4), 157–160

8. Beeson, R. C. (1992). Restricting overhead irrigation to dawn limits growth in container-grown ornamentals. *Hort. Science,* 27, 996–999.

9. Biernbaum, J. A. (1993). Subirrigation could make environmental and economical sense for your greenhouse. *PPGA News,* p. 2–14.

10. Bouma, J. R, Brown, B., & Rao, S. C. (2003). *Movement of water: Basics of Soil Water Relationships – Part III.* UFAS extension fact sheet SL-39, Florida.

11. Bravdo, B., & Proebsting, EL. (1993). Use of drip irrigation in orchards. *Hort. Techn.* 3 (1), 44–49.

12. Brian, R. R. (2001). *Egyptian Potato Exports to the EU.* FAO statistics. p. 115–130.

13. Burton, W. G. (1984). *The potato.* Chapman and Hall, London. 319 pp.

14. Dave, D. (2003). Egypt and the potato tuber moth. Dave Douches at MSU. 280 pp.

15. David, T. (2006). *Milestone Report No 1 Sustainable Horticultural Irrigation Project* (SHIP). University of Western Australia. 120 pp.

16. Dole, J. M. (1993). Water and fertilizer rate reduction. *Greenhouse Grower,* 11(13), 24–28.

17. Dole, J. M. (1994). Comparing poinsettia irrigation methods. *The Poinsettia,* 10, 4–9.

18. Dole, J. M., Cole, J. C., & von Broembsen, S. L. (1994). Growth of poinsettias, nutrient leaching, and water-use efficiency respond to irrigation methods. *HortScience,* 29, 858–864.

19. Doorenbos, J., & Pruitt, W. O. (1977). Guidelines for predicting crop water requirements FAO, *Irrigation and Drainage Paper No. 24. FAO,* Rome, 270 pp.

20. Dubbois, M., Gilles, K. A., Hamilton, J. K., Rebers, P. A., & Smith, F. (1956). Colorimetric methods for determination of sugars and related substances. *Analytical Chem.,* 28, 250–356.

21. El-Adi, A. M. (2000). Effect of irrigation and fertilization methods on pea production. *Misr. J. Ag. Eng.,* 17(3), 450–468.

22. El-Berry, A. M., & Bakeer, G. A. (1995). *Using Subsurface Drip Irrigation Technology for Vegetables' Production.* Guide bulletin, T.T.C. Agriculture Faculty, Cairo University. 12 pp.

23. El-Gindy, A. M., & Abdel Aziz, A. A. (2001). Maximizing water use efficiency of maize crop in sandy soils. *Arab Univ. J. Agric. Sci., Ain Shams Univ.,* Cairo, Egypt, 11(1), 439–452.

24. Elliott, G. C. (1992). A pulsed subirrigation system for small pots. *HortScience,* 27, 71–72.

25. El-Meseery, A. A. (2003). Effect of different drip irrigation systems on maize yield in sandy soil. *The 11th Annual Conference of Misr Society of Agr. Eng.*, 15–16 Oct: 576–594

26. Eltawil, M. D., & Singhal, O. (2006). *Potato Storage Technology and Store Design Aspects.* Agricultural Engineering International: the CIGR E-journal. Invited Overview No.11, Vol. VIII.

27. Elwin, A. R. (1997). *Irrigation Guide. National Engineering Handbook. Natural Resources Conservation Service.* United States Department of Agriculture. pp. 652.

28. Eric, S., David, S., Robert, H. (2004). *To Pulse or Not to Pulse Drip Irrigation That is the Question.* UF/IFAS. Florida, USA.

29. Ertek, A., Sensoy, S., Yýldýz, M., & Kabay, T. (2002). Estimation of the most suitable irrigation frequencies and quantities in eggplant grown in greenhouse condition by using free pan evaporation coefficient. *K.S. Univ. Life Sci. Eng. J.*, 5(2), 57–67.

30. Feng-Xin, W., Kang, Y., & Liu, S. (2006). Effects of drip irrigation frequency on soil wetting pattern and potato growth in North China Plain. Center for Agricultural Water Research in China. China Agricultural University, *Agricultural Water Management*, 79, 248–264.

31. Freeman, B. M., Blackwell, J., & Garzoli, K. V. (1976). Irrigation frequency and total water application with trickle and furrow systems. *Agric. Water Manage.*, 1, 21–31.

32. Georgiev, D. (1997). New low-pressure system for pulse drip irrigation. *International Water and Irrigation Review,* 17, 1–19.

33. Georgiev, D., & Conley, A. H. (1996). Construction and hydraulic testing of an open hydraulically operated tank for pulse drip irrigation. Volume 1-E. *Transactions of the 16th International Congress on Irrigation and Drainage*, 1E, 167–176.

34. Gladis, Z. (2005). *Irrigation Management Options for Containerized-Grown Nursery Crops.* www.rcre.rutgers.edu.

35. Goodwin, I., & Boland, A. M. (2001). *Scheduling deficit irrigation of fruit trees for optimizing water use efficiency.* Department of Natural Resources and Environment, Institute of Sustainable Irrigated Agriculture, Tatura, Australia, p. 11–25.

36. Goodwin, I. (2005). *Water nutrient and salt balance simulations. Stage 1 Open Hydroponics: Risks and Opportunities.* National Program for Sustainable Irrigation report. 215–220 pp.

37. Goyal, Megh, R. (ed.), (2015). *Research Advances in Sustainable Micro Irrigation, volumes 1 to 10.* Oakville, ON, Canada: Apple Academic Press Inc.

38. University of Guelph (2002). *Organic Conference Organic Agriculture & the Farm Economy.* January 24–27 at University of Guelph Canada. 10–20 pp.

39. Hanson, B. R., May, D. M., & Schwankl, L. J. (2003). Effect of irrigation frequency on subsurface drip irrigated vegetables. Department of Land, Air and Water Resources, University of California, Davis, CA 95616, USA. *HortTechnology.* 13, 115–120

40. Helen, R. (2007). *Citrus Irrigation.* Department of Agriculture and Food, Waroona. State of Western Australia. http://www.Agric.wa.gov.au.

41. Helga, W., Yussefi, M., & Sorensen, N. (2008). *World of Organic Agriculture.* Statistics and Emerging Trends 2008. Earthscan, London, UK, 600 pp.

42. Helmy, M. A., Gomaa, S. M., Khalifa, E. M., & Helal, A. M. (2000). Production of corn and sunflower under conditions of drip and furrow irrigation with reuse of agricultural drainage water. *Misr. J. Ag. Eng.*, 17(1), 125–147.

43. Hillel, D. (1987). *The Efficient Use of Water in Irrigation*. World Bank Technical Paper, ISSN 0253-7494. 69–74 pp.

44. Jackson, R. C., & Kay, M. G. (1987). Use of pulse irrigation for reducing clogging problems in trickle emitters. *J. Agri. Eng. Res.,* 37, 223–227.

45. James, L. G. (1988). *Principles of Farm Irrigation System Design*. John Willey & Sons. Inc., pp. 73, 152–153, 350–351.

46. Keller, J., & Karmeli, D. (1975). Trickle irrigation design parameters. *Trans. ASAE,* 17(4), 678–684.

47. Kenig, E., Mor, E., Oronand, G., & Lamm, F. R. (1995). Pulsing micro irrigation for optimal water use and control in the soil. *Proceedings of the Fifth International Micro Irrigation Congress*, Orlando, Florida, USA, 2–6 April, ASAE, p. 615–620.

48. Kirda, C. (2000). *Deficit Irrigation Scheduling Based on Plant Growth Stages Showing Water Stress Tolerance*. Cukuroya University, Adana, Turkey, 300 pp.

49. Kirsten, B. (2007). *Organic Agriculture and Food Utilization*. Newcastle University, United Kingdom Kirsten.

50. Kolganov, A. V. (1995). Water -Saving technology of mist irrigation of maize for silage under the condition of light chestnut soils of the Lower Volga area. *Synopsis of the Thesis*, Moscow, 180 pp.

51. Kolganov, A. V., Nosenko, V. F. (2000). *Pulse Micro Irrigation Technology in Russia.* 6th International Micro irrigation Congress, Cape Town, South Africa, 22–27 Oct. pp. 1–8

52. Kushnirenko, M. D., Bykov, V. G., & Kurchatova, G. P. (1979). Water Exchange and Productivity of Apple Trees Under Synchronous Pulse Sprinkling. Kishinev, Shtiintsa, 146 pp.

53. Lieth, J. H. (1994). Controlling poinsettia irrigation based on moisture tension. The Poinseia, 10, 2–4.

54. Lina, A. L. (2004). *Report on Organic Agriculture in the Mediterranean Area*. CIHEAM (Centre International de Hautes Etudes Agronomi-ques Méditerranéennes) Bari p. 116, Série B: N. 50. Italia.

55. Liven, P. C., & Van Rooyen, F. C. (1979). The effect of discharge rate and intermittent water application by point-source irrigation on the soil moisture distribution pattern. *Soil Sci. Amer. J.,* 43, 8–5

56. Metin, S. S., Yazar, A., & Eker, S. (2005). *Effect of Drip Irrigation Regimes on Yield and Quality of Field Grown Bell Pepper.* Department of Water Management, Rural Services Tarsus Research Institute, Tarsus, Mersin, Turkey, 320 pp.

57. Mohamed, Abel-Rahman (2003). Study the effect of irrigation systems with different moisture content in soil on potato production and water use efficiency. *Misr. J.Ag. Eng.,* 15, 151–168.

58. Nautiyal, P. C., Joshi, Y. C., & Dayal, D. (2000). *Response of Groundnut to Deficit Irrigation During Vegetative Growth*. National Research Centre for Groundnut, Gujarat, India, 411 pp.

59. Netafim Landscape Division (2008). *Dripline Irrigation Manual*, and *Netafim_Landscape_Dripline_Manual.pdf*

60. Nosenko, V. F., Balabanand, E. I., & Landes, G. A. (1991). Aspects of agrobiological and environmental assessment of irrigation technologies with different volumes of water delivery. *Land Reclamation and Water Management. Central Bureau of Scientific and Technical Information*, Moscow, p. 71.

61. Oron, G. (1981). Simulation of water flow in the soil under sub-surface trickle irrigation with water uptake by roots. *Agric. Water Management,* 3, 179–193.

62. Qingwu, X., Zhuc, Z., Musickb, J. T., Stewartd, B. A., & Dusekb, D. A. (2006). Physiological mechanisms contributing to the increased water-use efficiency in winter wheat under deficit irrigation. *J. Plant Physiol.,* pp. 154–164 www.elsevier.de/jplphb, USA.

63. Rain Bird, (2008). ESP-LX Modular Controller Installation, Programming and Operation Guide, www.rainbird.com.

64. Rizk, E. K. (2007). Irrigation scheduling and environmental stress coefficient of kidney bean under some irrigation systems in North Sinai. *Egypt. J. Appl. Sci.,* 22(11), 286–296.

65. RO-DRIP® User Manual, (2001). Roberts Irrigation Products, Inc. 700 Rancheros Drive San Marcos, CA 92069–3007, USA. www.robertsirrigation.com.

66. Scott, C. (2000). Pulse Irrigation. *Water Savings Indiana Flower Growers Association.* Vol. 14, No.1. Cooperating with The Department of Horticulture and Landscape Architecture Cooperative Extension Service Purdue University West Lafayette, 120 pp.

67. Sean, M. (1996). *Pulse Irrigation Strategies for Greenhouse Production.* MSc thesis for Department of Horticulture and Landscape Architecture, Colorado State Univ. Fort Collins, Colorado, 220 pp.

68. Segal, E., Ben-Gal, A., & Shani, U. (2000). Water availability and yield response to high-frequency Micro irrigation in sunflowers. *6th International Micro irrigation Congress. Micro irrigation Technology for Developing Agriculture,* South Africa, 22–27. E-mail alonben-gal@rd.ardom.co.il

69. Sharmasarkar, F. C., Sharmasarkar, S., Miller, S. D., Vance, G. F., & Zhang, R. (2001). Assessment of drip irrigation and flood irrigation on water and fertilizer use efficiencies for sugarbeets. *Agric. Water Manage.,* 46, 241–251.

70. Shock, C., Flock, R., Eldredge, E., Pereira, A., & Jensen, L. (2006). *Drip Irrigation Guide for Potatoes in the Treasure Valley.* EM 8912-E, 442 pp.

71. Shock, C. C., Flock, R. J., Eldredge, E. P., Pereira, A. B., & Jensen, L. B. (2006). Successful Potato Irrigation Scheduling. *Oregon State University Extension Service Publication* EM 8911-E.

72. Snedecor, G. W. & Cochran, W. G. (1982). *Statistical Methods.* 7th ed., Iowa State Univ. Press, Towa, USA, 511 pp.

73. Solomon, K. (1999). *Irrigation Equipment Performance Report.* Center for Irrigation Technology. Fresno, CA, 342 pp.

74. Steve, D. (2001). *Resource Guide to Organic and Sustainable Vegetable Production.* National Sustainable Agriculture Information Service PO Box (3657). Fayetteville, AR 72702 USA.

75. Steven, F. (2006). *Economic and Environmentally Sustainable Citrus Production.* Bunbury District Office, Department of Agriculture Western Australia. E-mail: amccarthy@agric.wa.govau

76. Steyn, J. M., Duplessis, H. F., Fourie, P., & Roos, T. (2005). *Irrigation Scheduling of Drip Irrigation Potatoes.* Northern Province Dept. of Agriculture, P O Box 243, Pietersburg, 0900 South Africa. E- mail: marap@vopi.agric.za

77. Steyn, J. M., Plessis, H. F., & Fourie, P. (1998). *Response of Potato Genotypes to Different Irrigation Water Regimes.* Report of the Water Research Commission by the ARC-Roodeplaat Vegetable and Ornamental Plant Institute. WRC Report No. 389/1/98, Pretoria, South Africa, 321 pp.

78. Steyn, J. M., Plessis, H. F., Fourie, P., & Roos, T. (2000). Irrigation scheduling of drip irrigation potatoes. *6th International Micro Irrigation Congress. 'Micro irrigation Technology for Developing Agriculture.* E-mail: marap@vopi.agric.za

79. Sultan, W. M. (1995). *A Study on Modern Technique of Irrigation and Fertigation in Greenhouse.* M.Sc. in Agriculture Science, Al- Azher University, 160 pp.

80. Thompson, C. (2001). Irrigation and intensive large-scale crop management. *Proceedings of the 10th Australian Agronomy Conference*, Hobart., 121–144.

81. Van loon, C. D. (1981). The effect of water stress on potato growth, development and yield. *Am. Potato J.*, 51–69 pp.

82. Worth, B., & Xin, J. (1983). *Farm Mechanization for Profit.* Granada Publishing Co., UK, 531 pp.

83. Yardeni, A. (1989). Pulsation for better drip irrigation. *Water and Irrigation Review*, 9: 8–12.

84. Yelanich, M. V., & Biernbaum, J. A. (1990). Effect of fertilizer concentration and method of application on media nutrient concentration, nitrogen runoff and growth of *Euphorbia pulcherrima* 'V-17 Glory.' *Acta Hort.*, 272, 185–189.

85. Zin El-Abedin, T. K. (2006). Effect of pulse drip irrigation on soil moisture distribution and maize production in clay soil. *The 14th Annual Conference of Misr Society of Agr. Eng.*, 22 Nov., pp. 1058–1076.

APPENDIX I

Estimation of Actual Irrigation Requirements for Potato Crop According to the Meteorological Data of the Central Laboratory for Agricultural Climate

Date	Eto (mm/ day)	Kc	Etc (mm/ day)	WR (mm/day)	WR (L/plant/ day)	WR (m³/fed./ day)	WRa (m³/fed./ day)
Feb., 2	3.64	0.50	1.82	0.33	0.07	1.40	—
Feb., 3	3.67	0.52	1.91	0.35	0.07	1.47	—
Feb., 4	3.71	0.53	1.97	0.36	0.08	1.51	—
Feb., 5	3.74	0.55	2.06	0.38	0.08	1.58	—
Feb., 6	3.78	0.56	2.12	0.39	0.08	1.63	—
Feb., 7	3.81	0.58	2.21	0.40	0.08	1.70	9.3
Feb., 8	3.85	0.59	2.27	0.42	0.09	1.75	—
Feb., 9	3.89	0.61	2.37	0.43	0.09	1.82	—
Feb., 10	3.92	0.63	2.47	0.45	0.09	1.90	—
Feb., 11	3.96	0.64	2.53	0.46	0.10	1.95	—

Date	Eto (mm/ day)	Kc	Etc (mm/ day)	WR (mm/day)	WR (L/plant/ day)	WR (m³/fed./ day)	WRa (m³/fed./ day)
Feb., 12	3.99	0.66	2.63	0.48	0.10	2.02	—
Feb., 13	4.03	0.67	2.70	0.49	0.10	2.07	11.5
Feb., 14	4.06	0.69	2.80	0.51	0.11	2.15	—
Feb., 15	4.10	0.71	2.91	0.53	0.11	2.24	—
Feb., 16	4.14	0.72	2.98	0.55	0.11	2.29	—
Feb., 17	4.17	0.74	3.09	0.56	0.12	2.37	—
Feb., 18	4.21	0.75	3.16	0.58	0.12	2.43	—
Feb., 19	4.24	0.77	3.26	0.60	0.13	2.51	4.0
Feb., 20	4.28	0.78	3.34	0.92	0.19	3.85	4.1
Feb., 21	4.31	0.80	3.45	0.95	0.20	3.97	4.1
Feb., 22	4.35	0.80	3.48	0.96	0.20	4.01	4.1
Feb., 23	4.39	0.81	3.56	0.98	0.20	4.10	4.1
Feb., 24	4.42	0.82	3.62	0.99	0.21	4.18	4.1
Feb., 25	4.48	0.83	3.72	1.02	0.21	4.29	4.1
Feb., 26	4.49	0.84	3.77	1.38	0.29	5.80	6.1
Feb., 27	4.53	0.85	3.85	1.41	0.30	5.92	6.1
Feb., 28	4.56	0.86	3.92	1.44	0.30	6.03	6.1
Mar., 1	4.60	0.87	4.00	1.46	0.31	6.15	6.1
Mar., 2	4.65	0.88	4.09	1.50	0.31	6.29	6.1
Mar., 3	4.70	0.89	4.18	1.53	0.32	6.43	6.1
Mar., 4	4.75	0.90	4.28	1.96	0.41	8.21	8.7
Mar., 5	4.80	0.91	4.37	2.00	0.42	8.39	8.7
Mar., 6	4.85	0.92	4.46	2.04	0.43	8.57	8.7
Mar., 7	4.90	0.93	4.56	2.08	0.44	8.75	8.7
Mar., 8	4.95	0.94	4.65	2.13	0.45	8.94	8.7
Mar., 9	5.00	0.96	4.80	2.20	0.46	9.22	8.7
Mar., 10	5.05	0.97	4.90	2.69	0.56	11.29	11.9
Mar., 11	5.10	0.98	5.00	2.74	0.58	11.52	11.9
Mar., 12	5.15	0.99	5.10	2.80	0.59	11.75	11.9
Mar., 13	5.20	1.00	5.20	2.85	0.60	11.99	11.9
Mar., 14	5.25	1.01	5.30	2.91	0.61	12.22	11.9
Mar., 15	5.30	1.02	5.41	2.97	0.62	12.46	11.9

Date	Eto (mm/day)	Kc	Etc (mm/day)	WR (mm/day)	WR (L/plant/day)	WR (m³/fed./day)	WRa (m³/fed./day)
Mar., 16	5.35	1.03	5.51	4.54	0.95	19.06	20.0
Mar., 17	5.40	1.04	5.62	4.62	0.97	19.42	20.0
Mar., 18	5.45	1.05	5.72	4.71	0.99	19.79	20.0
Mar., 19	5.50	1.06	5.83	4.80	1.01	20.16	20.0
Mar., 20	5.55	1.07	5.94	4.89	1.03	20.54	20.0
Mar., 21	5.60	1.08	6.05	4.98	1.05	20.91	20.0
Mar., 22	5.65	1.09	6.16	6.76	1.42	28.40	29.1
Mar., 23	5.70	1.10	6.27	6.88	1.45	28.91	29.1
Mar., 24	5.75	1.10	6.33	6.94	1.46	29.16	29.1
Mar., 25	5.80	1.09	6.32	6.94	1.46	29.15	29.1
Mar., 26	5.85	1.09	6.38	7.00	1.47	29.40	29.1
Mar., 27	5.90	1.08	6.37	7.00	1.47	29.38	29.1
Mar., 28	5.95	1.07	6.37	8.15	1.71	34.25	34.5
Mar., 29	6.00	1.07	6.42	8.22	1.73	34.53	34.5
Mar., 30	6.05	1.06	6.41	8.21	1.72	34.50	34.5
Mar., 31	6.10	1.06	6.47	8.28	1.74	34.78	34.5
Apr., 1	6.10	1.05	6.41	8.20	1.72	34.45	34.5
Apr., 2	6.16	1.04	6.41	8.21	1.72	34.46	34.5
Apr., 3	6.22	1.04	6.47	8.88	1.86	37.28	37.5
Apr., 4	6.29	1.03	6.48	8.89	1.87	37.34	37.5
Apr., 5	6.35	1.02	6.48	8.89	1.87	37.33	37.5
Apr., 6	6.41	1.02	6.54	8.97	1.88	37.68	37.5
Apr., 7	6.47	1.01	6.53	8.97	1.88	37.66	37.5
Apr., 8	6.53	1.00	6.53	8.96	1.88	37.64	37.5
Apr., 9	6.60	1.00	6.60	9.66	2.03	40.57	40.7
Apr., 10	6.66	0.99	6.59	9.65	2.03	40.53	40.7
Apr., 11	6.72	0.98	6.59	9.64	2.02	40.49	40.7
Apr., 12	6.78	0.98	6.64	9.73	2.04	40.85	40.7
Apr., 13	6.84	0.97	6.63	9.71	2.04	40.79	40.7
Apr., 14	6.91	0.97	6.70	9.81	2.06	41.21	40.7
Apr., 15	6.97	0.96	6.69	10.41	2.19	43.71	43.9
Apr., 16	7.03	0.95	6.68	10.39	2.18	43.62	43.9
Apr., 17	7.09	0.95	6.74	10.48	2.20	44.00	43.9

Date	Eto (mm/day)	Kc	Etc (mm/day)	WR (mm/day)	WR (L/plant/day)	WR (m³/fed./day)	WRa (m³/fed./day)
Apr., 18	7.16	0.94	6.73	10.47	2.20	43.96	43.9
Apr., 19	7.22	0.93	6.71	10.44	2.19	43.86	43.9
Apr., 20	7.28	0.93	6.77	10.53	2.21	44.22	43.9
Apr., 21	7.34	0.92	6.75	11.12	2.34	46.70	46.8
Apr., 22	7.40	0.91	6.73	11.09	2.33	46.57	46.8
Apr., 23	7.47	0.91	6.80	11.19	2.35	47.01	46.8
Apr., 24	7.53	0.90	6.78	11.16	2.34	46.87	46.8
Apr., 25	7.59	0.89	6.76	11.12	2.34	46.72	46.8
Apr., 26	7.65	0.89	6.81	11.21	2.35	47.09	46.8
Apr., 27	7.71	0.88	6.78	11.79	2.48	49.53	49.7
Apr., 28	7.78	0.88	6.85	11.90	2.50	49.98	49.7
Apr., 29	7.84	0.87	6.82	11.86	2.49	49.79	49.7
Apr., 30	7.90	0.86	6.79	11.81	2.48	49.60	49.7
May, 1	7.90	0.86	6.79	11.81	2.48	49.60	49.7
May, 2	7.97	0.85	6.77	11.78	2.47	49.46	49.7
May, 3	8.05	0.85	6.84	12.52	2.63	52.58	52.8
May, 4	8.12	0.84	6.82	12.48	2.62	52.42	52.8
May, 5	8.19	0.84	6.88	12.59	2.64	52.87	52.8
May, 6	8.27	0.83	6.86	12.56	2.64	52.75	52.8
May, 7	8.34	0.83	6.92	12.67	2.66	53.19	52.8
May, 8	8.41	0.82	6.90	12.62	2.65	52.99	52.8
May, 9	8.49	0.81	6.88	12.58	2.64	52.85	53.0
May, 10	8.50	0.81	6.89	12.60	2.65	52.91	53.0
May, 11	8.63	0.80	6.90	12.63	2.65	53.05	53.0
May, 12	8.71	0.79	6.88	12.59	2.64	52.88	53.0
May, 13	8.78	0.79	6.94	12.69	2.67	53.30	53.0
May, 14	8.85	0.78	6.90	12.63	2.65	53.05	53.0
May, 15	8.93	0.78	6.97	11.47	2.41	48.17	42.6
May, 16	9.00	0.77	6.93	11.41	2.40	47.93	42.6
May, 17	9.07	0.76	6.89	11.35	2.38	47.67	42.6
May, 18	9.15	0.76	6.95	8.91	1.87	37.41	42.6
May, 19	9.22	0.75	6.92	8.86	1.86	37.20	42.6
May, 20	9.29	0.74	6.87	8.80	1.85	36.98	42.6

Date	Eto (mm/ day)	Kc	Etc (mm/ day)	WR (mm/day)	WR (L/plant/ day)	WR (m³/fed./ day)	WRa (m³/fed./ day)
May, 21	9.37	0.74	6.93	6.34	1.33	26.64	26.6
May, 22	9.44	0.73	6.89	6.30	1.32	26.48	26.6
May, 23	9.51	0.73	6.94	6.35	1.33	26.67	26.6
May, 24	9.59	0.72	6.90	6.32	1.33	26.53	26.6
May, 25	9.66	0.71	6.86	3.76	0.79	15.81	15.9
May, 26	9.73	0.71	6.91	3.79	0.80	15.93	15.9
May, 27	9.81	0.70	6.87	3.77	0.79	15.83	15.9
Applied water per season (m³)							3076
Amount of applied water before planting (m³)							400
Actual water requirements per season (m³) = WRa							3476

APPENDIX II

Typical Drip Irrigation Systems

CHAPTER 8

IMPACT OF POLYETHYLENE MULCH ON MICRO IRRIGATED CABBAGE

M. U. KALE, S. B. WADATKAR, S. M. GHAWADE,
J. N. LOKHANDE, and A. S. TALOKAR

CONTENTS

8.1   INTRODUCTION

Increasing population, growing urbanization, and rapid industrialization combined with the need for raising agricultural production generates competing claims for water from domestic, industrial and agricultural sector. There is a growing perception of a sense of an impending water crisis in India. In the past few decades, over exploitation of surface and groundwater has resulted in water scarcity in some regions. Irrigation, being the major water user, its share in the total freshwater demand is bound to decrease from the present 83% to 74% due to more pressing and competing demands from other sectors by 2025 AD [10], and the country will face water scarcity if adequate and sustainable water management initiatives are not implemented.

Water is the major limiting factor for crop diversification and production. Reductions in water availability for irrigation use increases the importance of implementation of water conservation practices in agriculture. Optimizing water use is an economic and environmental concern for agricultural producers. Therefore, irrigation is required for profitable production. There is a need to double annual food grain production from the present 210 million tons to 420 million tons within next 10 years. Since land is a shrinking resource for agriculture, the pathway for achieving this goal has to be higher productivity per unit of arable land and water. Thus, the objective of irrigation in the present era is not only to provide supplementary water for crop production but also to increase crop per drop of water.

Improper irrigation management practices not only waste expensive and scarce water resources but also decreases crop yield, quality, water use efficiency and economic return. It might lead to water logging and salinity, as well, which can be partly corrected by expensive drainage system.

Drip irrigation applies small quantity of water at frequent intervals, and it is superior in terms of water productivity and fruit quality [2, 3, 6]. It also makes possible the application of fertilizers and other chemicals along with irrigation water to match the plant requirements at various growth stages. Water application efficiency in the drip irrigation is higher than other methods of irrigation. Drip irrigation system accounts 40–50% of water saving and also results in 15–20% more yield coupled with 30% saving of fertilizer and labor compared with surface irrigation method. Only limitation of this system is initial high investment.

Crop productivity was enhanced with use of mulches [1, 4, 11]. Use of mulch has many benefits like increase in soil temperature especially in early spring, reduction in weeds, moisture conservation and higher crop yields [7, 9]. Mulching also helps in improving the soil structure, soil fertility and soil biological regime. Use of drip irrigation in combination with mulch can increase the crop yield significantly over drip irrigation alone to the tune of 20–30%.

Cabbage is one of the most common cole crops, which thrives best in cool weather. Nutritional value of 100 g of edible portion of cabbage contains: 1.8 g protein, 0.1 g fat, 4.6 g carbohydrate, 0.6 g mineral, 29 mg calcium, 0.8 mg iron, 14.1 mg sodium, besides enriched in Vitamins A and C. Cabbage has a number of varieties both wild as well as cultivable. Average annual production of cabbage in India is 8,395 thousand tons and in Maharashtra it is 421 thousand tons [8]. Productivity of cabbage in India, Maharashtra and Vidarbha region is 22.04, 21.00, and 20.1 t/ha, respectively.

Irrigation scheduling plays a vital role in crop productivity. Full irrigation scheduling (ETo) is observed as the best by the researchers for drip irrigated crops under field conditions, while in case of limited water supply, adopting deficit irrigation strategies could be an alternative for irrigation scheduling of crop under semi-arid region. There is a need of local experimental verification of effects of mulching with drip irrigation under different irrigation scheduling to improve the water productivity.

This chapter focuses on the feasibility of use of different colored mulches along with drip irrigation in increasing the water productivity for cabbage production in semi-arid area.

8.2   MATERIALS AND METHODS

8.2.1   EXPERIMENTAL SITE

The experiment was conducted at research farm of Department of Irrigation and Drainage Engineering, Dr. PDKV, Akola, situated in Western Vidarbha region of Maharashtra State and comes under subtropical zone. It is situated at an altitude of 307.4 m above mean sea level at the intersection of 20°42′ North latitude and 77°02′ East longitude. Average annual precipitation of Akola is 760 mm.

The minimum temperature over the period of study varied from 7.4 to 21.5°C whereas maximum temperature varied from 24.8 to 33.6°C. Mean daily relative humidity, daily sunshine hours, wind speed and evaporation ranged between 45 to 89%, 0.2 to 9.0 h, 0.03 to 1.61 m.s^{-1} and 2.2 to 9.0 mm.day^{-1}, respectively.

8.2.2   DATA COLLECTION

8.2.2.1   Soil Characterization

Soil profile study (Figure 8.1) was carried out to characterize the soil at experimental plot. Soil samples were collected from a pit, dug at the center of field, from the depths of 0–18 cm, 18–42 cm and 42–68 cm under supervision of soil physicist. Soil samples were tested for physio-chemical analysis in the laboratory and results are presented in Tables 8.1 and 8.2.

8.2.2.2   Water Retention Properties of Soil

Water retention of soil at 0, 0.33, 3, 5 and 15 bars was determined using pressure plate apparatus. The soil moisture retention curves were developed using these observations (Figure 8.2). The porosity of soil was determined using porosity cup. It is estimated as 0.51, 0.51, 0.40 cm^3cm^{-3}, for soil horizon AP, A and C, respectively.

8.2.2.3   Water Source and Its Quality

The source of water was water distribution network of university. The water was analyzed for its suitability for irrigation (Table 8.3).

FIGURE 8.1   Soil profile at experimental site.

TABLE 8.1   Soil Characteristics at Experimental Site

Soil Horizon	Soil depth	Sand	Silt	Clay	Textural class	Bulk density	Water retention at		Saturated moisture content
							0.33 bar FC	15 bar PWP	
	cm	%			—	g.cm^{-3}	cm^3 cm^{-3}		cm^3 cm^{-3}
Ap	0–18	10.2	31	58.8	Clay	1.25	0.34	0.18	0.51
A	18–42	9.6	30.2	60.2	Clay	1.27	0.35	0.20	0.51
C	42–68	50.3	12.4	37.3	Sandy clay	1.35	0.26	0.15	0.40

TABLE 8.2   Chemical Composition of Soil

Particulars	Soil depth, cm			Analytical method
	0–18	18–42	42–68	
Organic carbon (%)	0.44	0.42	0.38	Walkly and black method
Available nitrogen (kgha^{-1})	281.08	270.19	265.93	Alkaline potassium permagnate method
Available P$_2$O$_5$ (kgha^{-1})	16	18	17	Olsen's method
Available K$_2$O (kgha^{-1})	301	300	280	Flame photometer method
EC (dsm^{-1})	0.38	0.24	0.33	Conductivity bridge from 1:2.5 soil water ratio
pH	7.9	7.3	7.2	pH meter using 1: 2.5 soil water ratio

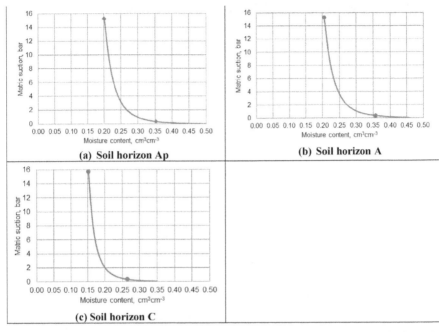

FIGURE 8.2 Soil moisture retention curves. (a) Soil horizon Ap; (b) Soil horizon A; (c) Soil horizon C.

TABLE 8.3 Chemical Analysis of Irrigation Water

Property	Value
pH	7.52
EC (ds.m^{-1})	0.46
HCO$_3$ (meq.lit^{-1})	1.30
Cl (meq.lit^{-1})	1.20
Ca + Mg (meq.lit^{-1})	1.60
Na (meq.lit^{-1})	3.95
K (meq.lit^{-1})	0.70
S.A.R.	4.90
R.S.C.	0.30

8.2.3 EXPERIMENTAL DESIGN AND TREATMENTS

The field experiment (Figure 8.3) was laid in randomized block design. The experiment consisted of four treatments with six replications. Following treatments were used:

T$_1$ Irrigation scheduling at 50% moisture depletion of available water capacity under black polyethylene mulch with drip irrigation

T$_2$ Irrigation scheduling at 50% moisture depletion of available water capacity under silver polyethylene mulch with drip irrigation

T$_3$ Irrigation scheduling at 50% moisture depletion of available water capacity with drip irrigation

T$_4$ Irrigation scheduling at 100% replenishment of evapotranspiration with drip irrigation

TABLE 8.4 Details of Experiment

Items	Specification
Name of the crop	Cabbage
Scientific name	*Brassica oleracea Var. capitata.*
Variety	Golden acre
Planting time	Rabi season
Design	RBD (Randomized Block Design)
Number of treatments	4
Number of replications	6
Crop spacing	45 cm x 50 cm
Plot size	3.5 m x 3.6 m
Number of plants/plot	42
Number of plots	24
Seed rate	500 g.ha^{-1}
Duration of crop	120 days
Mulch	Black and silver polyethylene mulch
Date of sowing	12–10–2013
Date of transplanting	26–11–2013
Date of last harvesting	19–02–2014
Fertilizer dose, NPK	150:50:0 kg ha^{-1}

FIGURE 8.3   Crop stand in the experimental plots.

8.2.4   EXPERIMENTAL SET UP

8.2.4.1   Land Preparation

Land was prepared with deep plowing followed by harrowing. Using broad bed furrow maker, the soil bed were raised to 15 cm, as shown in Figure 8.4.

8.2.4.2   Polyethylene Mulch

The polyethylene mulch film (silver and black) of 50 μm thickness was used to cover soil in the mulch treatments.

8.2.4.3   Laying of Gypsum Block Sensors in the Soil

Before installation, gypsum blocks were soaked in the water for two hours. Three holes of 30 cm depth were made in each plot in a zigzag fashion with the help of a screw auger (Figure 8.5), considering average 30 cm of root depth of cabbage. Tied paired gypsum blocks (Figure 8.5)

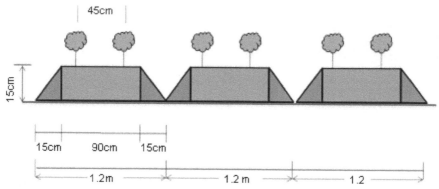

FIGURE 8.4   Dimensions of raised bed plot.

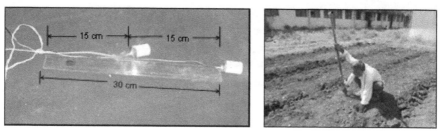

FIGURE 8.5   Installation of gypsum block in the soil at experimental plot.

were installed into the augured holes. The holes were then filled with soil resembling the original soil conditions.

8.2.4.4   Calibration of Moisture Meter

A table (chart, as well) was provided by the manufacturer showing relationship between moisture meter reading and soil moisture tension. Using developed soil moisture retention curves and chart provided by manufacturer, the moisture meter was calibrated for the soil at experimental site. An example of reading of a soil moisture sensor is shown in Figure 8.6.

Verification of Moisture Meter Reading

Before start of experiment, practically observed gravimetric soil moisture content and moisture content from soil moisture characteristic curve in reference to moisture meter reading were verified against each other.

FIGURE 8.6   Observation of moisture meter reading.

8.2.4.5   Irrigation System Set Up

The irrigation system consisted of: pump, suction pipe, delivery pipe, inline drip and other accessories, such as control valve, tee, elbow, coupling, reducer, G.T.O, etc. The water was conveyed to the experimental plot through installed pipeline at experimental site.

8.2.4.6   Fertilizer Application

The recommended dose of fertilizer in all plots was given through urea and phosphate at the rate of 150 kg and 50 kg per hectare, respectively. Half dose of N and full dose of P was given at the time of transplanting and remaining dose of N was given after one month of sowing.

8.2.4.7   Crop Harvesting

As all cabbage heads were not ready for harvest at one time, therefore these were harvested in stages based on the maturity of the heads and cumulative yield was recorded. Cabbage head was harvested on 20, 22, 27th January and 3, 11 and 19th February 2014, when the heads were firm and mature.

8.2.5 ESTIMATION OF IRRIGATION EFFICIENCY

8.2.5.1 Emission Uniformity

Irrigation efficiency is used to judge the performance of irrigation system. The application efficiency in the form of emission uniformity for inline drip system was calculated by using equation below [5]:

$$E_u = 100 \times \frac{q_n}{q_a} \qquad (1)$$

where, E_u = emission uniformity in %, q_n = average of lowest $1/4^{th}$ of emission point discharges for field data in lph, and q_a = average emission point discharge of test samples operated at the reference pressure head in lph.

8.2.5.2 Water Use Efficiency

Water use efficiency is the ratio of crop yield to the amount of irrigation water applied during crop period. It was calculated by using following equation:

$$E_{ui} = \frac{Y}{WR} \qquad (2)$$

where, E_{ui} = irrigation water use efficiency in $(100 \text{ kg}).ha^{-1}.cm^{-1}$, Y = crop yield in 100 kg, and WR = water requirement in ha-cm.

8.2.6 IRRIGATION SCHEDULING

Common irrigation was applied on 25th November 2013 to the experimental plot, to bring soil to field capacity. The first irrigation was given on 27th November 2013 up to field capacity in all treatments.

8.2.6.1 Irrigation Water Requirements for Treatments Based on Irrigation Scheduling At 50% Moisture Depletion of Available Water Capacity

Total available water (TAW) was determined using soil moisture constants of the soil and 30 cm root zone depth. Depth of irrigation water (IW) per irrigation was calculated considering 50% maximum allowable depletion, in T_1, T_2 and T_3. The soil moisture was observed every day with the moisture meter (Figure 8.6). Irrigation was provided, when moisture reduced to 50% of TAW. Irrigation water need at 50% depletion of AWC was calculated using following formula:

$$d = \sum_{i=1}^{n} \frac{(Yfci - Mbi)}{100} \times A_i \times D_i \tag{3}$$

where, d = net amount of water to be applied during an irrigation in cm, Mfci = field capacity of moisture content in the ith layer of the soil in percent, Mbi = moisture content before irrigation in the ith layer of the soil in percent, A_i = bulk density of the soil in the ith layer, D_i = depth of the ith soil layer within the root zone in cm, and n = number of soil layers in the root zone D.

8.2.6.1.1 Gross Irrigation Requirement

Gross irrigation requirement was calculated for experimental plot, considering appropriate losses using following relationship:

$$IR = \sum_{i=1}^{n} \frac{d}{E_{(application)}} \tag{4}$$

where, IR = gross irrigation requirement at the field head in cm, d = net irrigation requirement in cm, $E_{(application)}$ = water application efficiency in %, and n = number of irrigations in a season.

8.2.6.2   Irrigation Water Requirements for Treatments Based on Irrigation Scheduling At 100% Reference Evapotranspiration

Water requirement of the crop under drip irrigation at 100% replenishment of evapotranspiration was computed on daily basis based on pan evaporation by using following equation for treatment T_4:

$$ET_c = E_p \times K_c \times K_p \tag{5}$$

where, ET_c = crop evapotranspiration in mm.day^{-1}, K_c = crop coefficient, K_p = pan coefficient, and E_p = pan evaporation, mm.day^{-1}. The volume of water to be applied per treatment was calculated with following equation:

$$V = ET_c \times A \times N \qquad (6)$$

where, V = volume of water per treatment in liters, ET_c = crop evapotranspiration in mm.day^{-1}, A = area of one plot in m^2, and N = number of plots.

Values of pan coefficient and crop coefficients were taken from *Food and Agricultural Organization (FAO) Paper 56*. Water requirement of cabbage crop was estimated on daily basis over the crop period. Irrigation was scheduled on daily basis. The value of pan coefficient was assumed as 0.7 [12].

8.2.7   BIOMETRIC OBSERVATIONS

In order to observe growth, yield and quality of cabbage as affected by different irrigation methods, biometric observation (number of leaves/plant, plant height, leaf area) were observed at 15, 30, 45, 60 days after transplanting (DAT). Total yield of cabbage obtained for each harvesting was recorded and cumulated over crop period. Diameter of fully developed, cabbage head was also measured.

8.2.8   COST ECONOMICS OF CABBAGE PRODUCTION

Cost analysis (benefit–cost ratio) for all treatments was worked out to compare the net returns. For this purpose, the life period of polyvinyl chloride items was considered as 10 years and 25 years for G.I. items and motor pump set. Standard market rates were considered for each Item. Fixed cost, operating cost, net return and benefit–cost ratio for each treatment were worked out using standard formulae. Benefit–cost ratio was calculated using following relationship [13]:

$$\text{B:C ratioo} = \frac{\text{Net benefits per crop season}}{\text{Total cost of production of cabbage}} \qquad (7)$$

8.3 RESULTS AND DISCUSSION

8.3.1 IRRIGATION SCHEDULING

Depth of irrigation required in both cases of scheduling was determined. Table 8.5 indicates that irrigation of 26.4 mm was required to bring the soil to field capacity in case of treatment T_1, T_2, and T_3. The period after which irrigation is required depends on crop growth stage, soil type and climate. Therefore, it varied from treatment to treatment.

8.3.2 WATER REQUIREMENT OF CABBAGE

For each crop stage, irrigation water applied and effective rainfall received were summed to determine water requirement of cabbage in each crop growth stage, as influenced by different irrigation scheduling. Table 8.6 shows that maximum water was required by cabbage during mid season stage followed by crop development stage, initial and late season stage. Treatment T_2 required minimum water for cabbage production. Therefore maximum saving of water in Treatment T_2 took place during mid season stage.

8.3.3 WATER SAVING UNDER DIFFERENT TREATMENTS

Number of irrigations, water applied and water saving are presented in Table 8.7. It is observed that the highest number of irrigations was used in the treatments T_4, followed by T_3, T_1 and T_2. Irrigation applied under different treatments varied from 327.4 mm to 252.2 mm. Total water requirement of cabbage was highest (327.4 mm) under irrigation scheduling at 100% ET_c (T_4), followed by T_3 (305 mm), T_1 (278.6 mm) and T_2

TABLE 8.5 Irrigation Water Requirement of Cabbage

Parameter	Value
Total available water (TAW), %	13.92
Depth of water applied during each irrigation (IW), mm	
Treatment: T_1, T_2, and T_3	26.4
Treatment T_4	$= ET_c$

TABLE 8.6   Water Requirement for Each Growth Stage

Crop growth stage	Water requirement under different treatments, mm			
	T1	T2	T3	T4
Before transplanting	41.0	41.0	41.0	41.00
Initial	52.8	52.8	52.8	55.52
Crop development	52.8	52.8	79.2	88.96
Mid season	105.6	79.2	105.6	110.64
Late season	26.4	26.4	26.4	31.28
Seasonal water requirement	**278.6**	**252.2**	**305.0**	**327.4**

TABLE 8.7   Water Saving

Treatment	No. of Irrigations	Water applied, IW (mm)	Effective rainfall, ER (mm)	Total water applied, IW+ER (mm)	Water saving over control, (%)
T_1	9	278.6	0.0	278.6	14.91
T_2	8	252.2	0.0	252.2	22.97
T_3	10	305.0	0.0	305.0	6.84
T_4 (control)	81	327.4	0.0	327.4	0.00

(252.2 mm). The maximum water saving over control treatment was in treatment T_2 (22.97%), followed by T_1 (14.91%) and T_3 (6.84%).

8.3.4   SOIL MOISTURE DEPLETION

Soil moisture depletion study was undertaken in treatments T_1, T_2 and T_3 while soil moisture was not observed in case of T_4 as irrigation was scheduled daily equal to 100% crop evapotranspiration (ETc), a regular practice. Soil moisture was observed daily in treatments T_1, T_2 and T_3. Soil moisture depletion pattern is depicted in Figure 8.7. It is observed from Figure 8.6 (a, b and c) that the soil moisture was always maintained in allowable depletion regime.

8.3.5   PERFORMANCE OF CABBAGE

The statistical analysis of recorded biometric observations was carried out and is presented in Table 8.8.

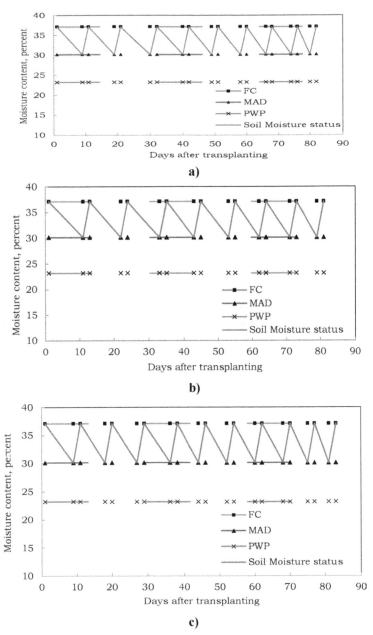

FIGURE 8.7 Soil moisture depletion pattern. (a) Treatment T_1; (b) Treatment T_2; (c) Treatment T_3.

TABLE 8.8   Statistical Analysis of Biometric Observations

| Treatment | Average value at 15, 30, 45 and 60 DAT | | | Diameter of cabbage head, cm | Yield (100 kg). ha⁻¹ |
	Plant height, cm	Number of leaves	Leaf area, cm²		
T₁	12.39	11.67	2.92	11.47	100.73
T₂	12.56	12.18	3.04	12.04	126.25
T₃	13.08	12.78	3.19	10.84	99.00
T₄ (control)	14.44	12.30	3.08	11.99	143.11
Mean	**13.56**	**12.24**	**3.06**	**11.77**	**122.62**
F-test	Sig.	Sig.	Sig.	Sig.	Sig.
SE (m) +	0.060	0.052	1.065	0.043	4.989
CD at 5%	0.184	0.158	3.208	0.132	15.030
CV (%)	0.899	0.857	2.762	0.745	8.139

It can be observed in Table 8.8 that all growth parameters were influenced significantly by different irrigation/mulch treatments. Treatments T_1 and T_3 showed cabbage head yield at par with each other, while T_2, and T_4 showed significantly highest yield as compared to treatments T_1 and T_3.

8.3.6   IRRIGATION EFFICIENCIES

Effects of irrigation scheduling and mulching on emission uniformity and water use efficiency (WUE) are presented in Table 8.9.

The emission uniformity of 92.5% for inline drip system was used in the experiment. It was observed that the maximum yield was obtained in treatment T_4 while minimum was recorded in treatment T_3. Maximum water was applied in treatments T_4 followed by T_3, T_1 and T_2. Table 8.9 shows that the treatment T_2 (drip + silver mulch, irrigation scheduling at 50% of AWC) recorded highest WUE followed by treatment T_4 (control), T_1 and T_2. Highest WUE was recorded in treatment T_2, which is due to the lowest water use.

8.3.7   COST ECONOMICS

Cost analysis of cabbage production under various treatments was carried out as shown in Tables 8.10 and 8.11. Fixed cost remained same

TABLE 8.9 Irrigation Efficiency Under Different Treatments

Treatment	Emission uniformity, %	Yield (100 kg). ha^{-1}	Consumptive use ha-cm	Water use efficiency (100 kg)/ha-cm
T$_1$	92.5	100.73	27.86	3.62
T$_2$		126.25	25.22	5.01
T$_3$		99.00	30.50	3.25
T$_4$ (Control)		143.11	32.74	4.37

TABLE 8.10 Total Cost For Cabbage Production (1.00 Rs. = 0.01566 US%)

Items	Treatments			
	T1	T2	T3	T6
Fixed cost, Rs ha^{-1}	5537	5537	5537	5537
Operating cost, Rs ha^{-1}	40081	40081	19011	19011
Total cost, Rs ha^{-1}	**45618**	**45618**	**24548**	**24548**

TABLE 8.11 Cost Analysis of Cabbage Production

Treatment	Yield of cabbage 100 kg/ha	Gross return*	Total cost (Rs ha^{-1})	Net return	BC ratio —
T$_1$	100.73	100730	45618	55112	1.21
T$_2$	126.25	126250	45618	80632	1.77
T$_3$	99.00	99000	24548	74452	3.03
T$_4$	143.11	143110	24548	118562	4.83

*The market rate of cabbage was Rs. 1000 per 100 kg or Rs. 10/kg.

in all treatments and only operating cost changed in each treatment. Therefore operating cost was calculated for each treatment separately. Total cost for cabbage production benefit–cost ratio for all treatments were estimated.

The BC ratio varied between 1.21 and 4.83. It was maximum for control treatment T$_4$, while it was minimum for treatment T$_1$. It was maximum for treatment T$_4$ due to less total cost as it does not include cost of mulch. Similarly, another non-mulch treatment T$_3$ also resulted in higher BC ratio (3.03). In mulch treatments, BC ratio was maximum for treatment T$_2$

followed by T_1. Under silver mulch, the BC ratio for cabbage production was highest as compared to that under black mulch.

8.4 CONCLUSIONS

On the basis of water use efficiency and benefit–cost ratio data, it was concluded that under water constraints, cabbage should be grown under silver mulch with drip irrigation having irrigation scheduling at 50% of AWC.

8.5 SUMMARY

The randomized block design field experiment having four treatments with six replications was conducted on cabbage at research farm of Department of Irrigation and Drainage Engineering, Dr. PDKV, Akola for its response under different irrigation/mulch treatments on water saving during November 2013 to February 2014. Irrigation was scheduled when moisture reduced to 50% of total available water TAW in case of first three treatments (T_1, T_2, and T_3), while in fourth treatment (T_4), daily irrigation equal to evapotranspiration was provided. Non-mulch drip irrigated treatment T_4 was the control. Silver and black colored polyethylene mulch of 50μ thickness was used in this study. Seasonal water requirement of cabbage was highest (327.4 mm) under irrigation scheduling at 100% ET_o (T_4, T_5, and T_6). It was lowest (252.2 mm) under irrigation scheduling at 50% moisture depletion of available water capacity under silver polyethylene mulch with drip irrigation. The highest saving of water over control treatment was achieved in T_2 (22.97%), followed by T_1 (14.91%) and T_3 (6.84%). It is also clear that the treatments with irrigation @ 50% depletion of AWC resulted in lower yield as compared to treatments with irrigation equal to 100% evapotranspiration. No specific pattern of yield in response to water/mulch was observed. Treatment T_2 (drip + silver mulch, irrigation @ 50% of AWC) gave highest water use efficiency followed by treatment T_6 (control), T_5, T_4, T_1 and T_3. The BC ratio was maximum for control treatment, for example, T_6, while it was minimum for treatment T_1. BC ratio for non-mulch treatments was higher than mulch treatments. It is observed that treatments with mulch and irrigation scheduling at 100% replenishment of evapotranspiration, resulted in higher BC ratio as compared to treatments with mulch and irrigation scheduling at 50% depletion of AWC. Under silver mulch, the BC ratio for cabbage production was higher as compared to that under black mulch in both cases of irrigation scheduling.

KEYWORDS

- **drip irrigation**
- **evapotranspiration**
- **gypsum block**
- **irrigation scheduling**
- **Mulch**
- **plastic mulch**
- **water requirement**
- **water use efficiency**

REFERENCES

1. Diaz-Perez, J. C. (2010). Bell Pepper (*Capsicum annum* L.) Grown on plastic film mulches: effects on crop microenvironment, physiological attributes, and fruit yield. *Hortscience*, 45(8), 1196–1204.
2. Himanshu, S. K., Kumar, S., Kumar, D., & Mokhtar, A. (2012). Effects of lateral spacing and irrigation scheduling on drip irrigated cabbage (*Brassica oleracea*) in a semi arid region of India. *Research J. Engineering Sciences*, 1(5), 1–6.
3. Jawale, P. V., & Dixit, M. S. (2007). Economic feasibility of drip irrigation system in Akola district. Unpublished BTech. Thesis, Dr. PDKV, Akola, pp. 22.
4. Jayasinghe, V., & Goonasekera, S. (1993). Influence of poly-ethylene film mulch on the yield of irrigated chili in the dry zone of Sri Lanka. *Tropical Agricultural Research*, 5, 42–49.
5. Keller, J., & Karmeli, D. (1974). Trickle irrigation design parameters. Transactions of the *American Society of Agricultural Engineers*, 17(4), 678–684.
6. Kumar, P., Sengar, S. S., & Agrawal, B. (2012). Effect of irrigation methods, levels and fertigation on cabbage. *Inter J Curr Trends Sci Tech*, 3(1), 37–41.
7. Patil, S. S., Kelkar, T. S., & Bhalerao, S. A. (2013). Mulching: A Soil and Water Conservation Practice. *Res. J. Agri. Forestry Sci.*, 1(3), 26–29.
8. National Horticulture Board, (2013). Statewise, area, production and productivity of cabbage. http://nhb.gov.in/area-pro/Indian%20Horticulture%202013.pdf

9. Rajablariani, H., Rafezi, R., & Hassankhan, F. (2012). Using colored plastic mulches in tomato (*Lycopersicon esculentum* L.) production. *International Conference on Agriculture and Animal Science*, 47, 12–16.

10. Swaminathan, M. S. (2006). Report of sub-committee on more crop and income per drop of water. *Advisory Council on Artificial Recharge of Ground Water Ministry of Water Resources, Government of India*: p. 1–61.

11. Tiwari, K. N., Mal, P. K., & Singh, A. (2003). Effect of drip irrigation on yield of cabbage (*Brassica oleracea* L. var. capitata) under mulch and non-mulch conditions. *Agricultural Water Management*, 58, 19–28.

12. Uttarwar, J. N. (1996). Effect of irrigation fertilizer levels on growth, yield, quality of Sugarcane. Unpublished M.Tech. Thesis, Dr. PDKV, Akola. p. 43–44.

13. Yasir, M., Muhammad, B. A., Muazzam, S., & Muhammad, A. (2011). Benefit–cost ratios of organic and inorganic wheat production: a case study of district Sheikhu-pura. *World Applied Sciences Journal*, 13(1), 175–180.

APPENDIX I   A TYPICAL CABBAGE FIELD UNDER DRIP IRRIGATION

CHAPTER 9

PERFORMANCE OF GARLIC UNDER DIFFERENT IRRIGATION SYSTEMS

M. U. KALE, S. B. WADATKAR, M. M. DESHMUKH, D. B. PALWE, and A. S. TALOKAR

CONTENTS

9.1 INTRODUCTION

Water is a precious natural resource, a basic human need and prime national asset. India will be a highly water stressed country 2020 onwards [7]. Population of India is expected to reach from 1027 million to 1930 million by 2025. Food grain requirement will be raised to 350 million tons by 2025 [1]. To meet this requirement there are two options either increase the gross area under irrigation or increase the water use efficiency. As water is becoming limiting resource, there is no scope to increase irrigated area by using additional water. Hence, only way to increase food production is by increasing the water use efficiency. For this purpose water saving and more yielding irrigation methods have to be used. Micro-irrigation systems satisfy this requirement.

The traditional surface irrigation methods are required to be replaced by modern water saving-more yielding irrigation methods like sprinkler, drip and micro irrigation. These are effective means of water saving. Micro-irrigation systems account 40–50% of water saving and also resulted in 15–20% more yield coupled with 30% saving of fertilizer and labor as compared with surface irrigation [6]. Only limitation of this system is high initial investment.

Garlic (*Allium sativum L.*) is second most important bulbous crop grown throughout the country. Garlic has higher nutritive value as compared to other bulbous crops. It is rich source of carbohydrates, proteins and phosphorus. Laboratory studies pointed out the protective values of garlic against heart disease, cancer and infectious disease. India rank second in area and production of garlic next to China. Production of garlic in India is 4.36×10^5 metric tons over total area of 1.07×10^5 hectare. The total productivity of garlic is 4.1 metric tons per hectare. Water requirement of garlic is 425 mm. Traditionally, garlic is irrigated with flood irrigation methods [4].

Various types of micro irrigation systems are now available in market. The selection of appropriate type of micro irrigation system is a critical task. Considering above mentioned aspects with view to grow more crop per drop of water, a field experiment aiming to check the performance of garlic under different irrigation systems was conducted.

9.2 MATERIALS AND METHODS

9.2.1 EXPERIMENTAL SITE

The experiment was conducted during *rabi* season of 2008 to 2011 at research farm of Department of Irrigation and Drainage Engineering, Dr. PDKV, Akola that is situated in Western Vidarbha region of Maharashtra State (India) and comes under subtropical zone. It is situated at an altitude of 307.415 m above mean sea level at the intersection of 20°42′ North latitude and 77°02′ East longitude. Average annual precipitation is 760 mm.

The climate of the area is semi-arid, characterized by three distinct seasons, namely summer being hot and dry from March to May, the warm and rainy monsoon from June to October and winter with mild cold from November to February. The maximum and minimum temperatures are 48.23°C and 22.05°C in summer and 32.88°C and 14.35°C in winter, respectively. Mechanical and chemical composition of soil at experimental site is presented in Tables 9.1 and 9.2. The soil moisture constants in the term of field capacity and permanent wilting point were determined using pressure plate apparatus and are given in Table 9.3. The source of water was analyzed to check its suitability for irrigation (Table 9.4).

TABLE 9.1   Mechanical Composition of Soil

Particulars	% composition
Sand	20.40%
Silt	40.59%
Clay	39.10%
Soil texture class	Clay loam

TABLE 9.2   Chemical Composition of Soil

Particulars	Soil depth, cm		
	10	20	30
Organic carbon	0.44	0.42	0.38
Available nitrogen (kg/ha)	281.08	270.19	265.93
Available P_2O_5 (kg/ha)	16	13	8
Available K_2O (kg/ha)	301	300	280
EC (ds/m)	0.44	0.47	0.48
pH	7.53	7.52	7.50

TABLE 9.3   Soil Moisture Constants at Experimental Plot

Particulars	Value
Depth of soil	30 cm
Field capacity	30.85%
Permanent wilting point	20.95%
Bulk density, g cc^{-1}	1.37

TABLE 9.4 Chemical Analysis of Irrigation Water

Particulars	Value
pH	7.52
EC, $ds.m^{-1}$	0.46
HCO_3, $meq.lit^{-1}$	1.30
Cl, $meq.lit^{-1}$	1.20
Ca + Mg, $meq.lit^{-1}$	1.60
Na, $meq.lit^{-1}$	3.95
K, $meq.lit^{-1}$	0.70
S.A.R.	4.90
R.S.C.	0.30

9.2.2 EXPERIMENTAL SET UP

The water was conveyed to online drip, micro sprinkler and check basin plots through installed pipeline at the experimental site. The irrigation system consisted of pump, suction pipe, delivery pipe, emitters, micro sprinklers and other accessories such as control valve, tee, elbow, coupling, reducer and G.T.O., etc. Details of experimental layout are as shown in Figure 9.1. The control head unit comprised of sand filter, screen filter, pressure gage, flow control valve and pressure regulating valve, etc.

The experiment was set in randomized block design having three treatments with seven replications. The details of treatments and experiment are given in Tables 9.5 and 9.6, respectively.

Plant height, number of leaves and yield were observed. After harvesting of garlic, it was cured in shed for about a month. Yield was observed before drying and after drying. The statistical analysis of the yield observations was carried out. Benefit–cost analysis was also worked out.

9.2.3 IRRIGATION MANAGEMENT

The amount of irrigation water required for all treatments to bring the soil to field capacity was calculated by using following equations [3].

FIGURE 9.1   Crop stand in experimental plots.

TABLE 9.5   Details of Treatments

Treatment	Irrigation system	Specification
T$_1$	Online drip	Dripper of 4 lph each, spacing 40 cm
T$_2$	Micro sprinkler	Single micro sprinkler of 64 lph discharge
T$_3$	Check basin	IW/CPE =1.2 with CPE = 40 -mm

TABLE 9.6   Details of Experiment

Particulars	Specification
Crop	Garlic
Scientific Name	*Allium sativum* L.
Variety	G-41
Experimental design	Randomized Block Design
Number of treatments	Three
Number of replications	Seven
Number of plots	21
Plot size	3.6 m x 3.6 m
Interspaces between plots	2 m
Crop Spacing	15 cm x 10 cm
Seed rate	500 kg/ha
Recommended fertilizer dose (N.P.K.)	100:50:50 kg/ha in two split
Date of sowing	17.11.2008; 24.11.2009; 19.11.2010
Date of harvesting	23.03.2009; 23.03.2010; 19.03.2011

$$d = \frac{Mfc - Mbi}{100} \times As \times Ds \qquad (1)$$

$$Q = d \times A \qquad (2)$$

where, Q = quantity of water required per plot in liter, Mfc = moisture content at field capacity in %, Mbi = moisture content before irrigation in %, As = apparent specific gravity in g/cc, Ds = depth of effective root zone in mm, d = net amount of water to be applied during an irrigation in mm, and A = area of plot, m^2.

9.2.3.1 Irrigation Scheduling

9.2.3.1.1 Micro Irrigation Treatments

The irrigation for micro irrigation treatments (inline drip and micro irrigation) was scheduled daily. After first common irrigation, the daily irrigation water requirement was determined using class A open pan evaporation, crop coefficient and pan coefficient. Daily evaporation data was obtained from the Meteorological Observatory, Department of Agronomy, Dr. PDKV, Akola during the period of investigation. The values of crop coefficient for different growth stages of garlic [2] are presented in Table 9.7.

The value of pan coefficient was taken as 0.7. The water requirement for garlic crop per day was calculated by using following equation [5]:

$$Q = A \times E_{pan} \times K_p \times K_c \qquad (3)$$

where, Q = quantity of water required in liters per day, A = area of plot in m^2, E_{pan} = pan evaporation in mm/ day, K_p = pan coefficient, and K_c = crop coefficient.

TABLE 9.7   Crop Coefficients for Garlic

Crop stages	Duration (days)	Crop coefficient
Initial	18	0.48
Crop development	60	0.94
Mid season	28	1.07
Late season	15	0.86

9.2.3.1.2 Check Basin Irrigation System

For check basin method, irrigation was scheduled at calculated pan evaporation (CPE = 40 mm with IW/CPE = 1.2). The amount of water (IW) required for each plot in the check basin irrigation system was calculated by using following equations:

$$\frac{IW}{CPE} = 1.2 \tag{4}$$

$$Q = IW \times A \tag{5}$$

where, IW = net amount of water to be applied during irrigation in mm, CPE = cumulative pan evaporation in mm/day, Q = quantity of water delivered in liters per plot, liters, and A = area of plot in m^2.

9.3 RESULTS AND DISCUSSION

9.3.1 IRRIGATION WATER REQUIREMENT

The amount of water applied to each plot under inline drip, micro irrigation and check basin irrigation method during various growth stages is presented in Table 9.8.

Table 9.8 clearly indicates that the maximum water was required by garlic during crop development stage compared to lowest during initial stage. During crop development stage, maximum vegetative growth took place and also it was longest duration stage. The mid-season stage is characterized with ceasing of vegetative growth and development of bulbs. Water requirement was decreased during the late season stage. During this stage, vegetative growth as well as bulb development completely ceases. Cumulative irrigation water requirement of garlic during cropping period under micro irrigation treatments was 47.49 ha-cm compared to 87.59 ha-cm under check basin irrigation treatment. Thus 45.78% water was saved through the use of micro-irrigation system over check basin irrigation system. Hence on an average, an additional area of 0.84 ha under garlic can be irrigated by using saved water adopting the micro irrigation system.

TABLE 9.8 Amount of Irrigation Water Applied to Garlic During Different Growth Stages

Growth stage	Duration (days)	Water applied, ha-cm						Average water applied, ha-cm	
		T1, T2			T3			T1, T2	T3
		2008–09	2009–10	2010–11	2008–09	2009–10	2010–11		
Just before sowing	—	5	6.3	6.72	5	6.3	6.72	6.01	6.01
Initial stage	18	3.21	2.22	2.41	9.6	9.59	4.80	2.61	8.00
Crop development stage	60	21.99	16.06	16.56	43.19	28.79	33.60	18.20	35.19
Mid season stage	28	13.63	14.51	13.77	28.79	23.99	24.00	13.97	25.59
Late season	15	4.70	7.49	7.90	9.6	14.39	14.40	6.70	12.80
Total		48.53	46.58	47.36	96.18	83.06	83.52	47.49	87.59
Water saved over traditional method								45.78%	

9.3.2 WATER USE EFFICIENCY (WUE)

The Table 9.9 indicates average weight of bulb, yield before drying, and yield after drying during 2008–09, 2009–10 and 2010–11. Yield before drying and after drying was maximum in online drip treatment compared to lowest yield in micro irrigation treatment. Yield obtained in online drip treatment was significantly higher than that in check basin treatment. While yield obtained in micro irrigation treatment was significantly lower than that in check basin treatment. Weight of bulb was found significantly higher in online drip irrigation treatment over that in check basin treatment.

Water use efficiency (WUE) for various irrigation treatments is presented in Table 9.10. WUE was maximum for online drip irrigation system followed by inline, micro irrigation and was lowest in case of check basin treatment.

TABLE 9.9 Pooled Analysis of Yield of Garlic Over the Period of Experiment (1.00 quintal, q = 100 kg)

Treatments	Yield of crop		
	Weight of Garlic before drying (q/ha)	Weight of Garlic after drying (q/ha)	Weight of bulb, (mg)
T1	72.71	45.80	17.8
T2	66.32	44.24	17.25
T3	49.6	30.86	16.6
Mean	**61.86**	**39.74**	**17.11**
F-test	Sig	Sig	Non Sig
SE (m) ±	1.20	0.72	0.48
CD at 5%	2.62	1.57	1.04
CV (%)	7.54	7.00	10.83
2008–09	60.97	40.11	16.75
2009–10	40.39	26.37	16.46
2010–11	84.23	52.72	18.13
Mean	**61.86**	**39.73**	**17.11**
F (test)	Sig	Sig	Sig
SE (m) ±	1.19	0.70	0.39
CD at 5%	2.42	1.43	0.79
CV (%)	8.65	7.94	10.26

TABLE 9.10   Water Use Efficiency Under Different Irrigation Treatments

Year	Water applied, ha-cm		Yield, 100 kg /ha		Water use efficiency (100 kg)/(ha-cm)	
	T1, T2	T3	T1	T3	T1	T3
2008–09	48.53	96.18	46.92	23.72	0.96	0.49
2009–10	46.58	83.06	28.70	24.90	0.61	0.53
2010–11	47.36	83.51	61.79	43.97	1.30	0.93
Average	**47.49**	**87.58**	**45.80**	**30.86**	**0.95**	**0.65**

9.3.3   BENEFIT–COST ANALYSIS

Benefit–cost (BC) analysis was carried out for garlic production. Market rate of garlic was assumed as Rs. 5000 per 100 kg. Total cost for micro-irrigation system was taken into consideration (Table 9.11). Benefit–cost ratio for various treatments are presented in Table 9.12.

Benefit–cost ratio for online irrigation treatment was maximum (1.96) followed by micro irrigation and check basin irrigation treatments. Hence, online drip irrigation system should be preferred for garlic production.

9.4   CONCLUSIONS

Inline drip irrigation system resulted in 20% higher yield and 46% water saving compared to traditional check basin irrigation method. Therefore, inline drip irrigation system is recommended to for garlic production.

9.5   SUMMARY

The field experiment in a randomized block design with three treatments and seven replications was conducted on garlic (*Allium sativum* L.) to study its performance under different irrigation methods during 2008 to 2011. The three treatments were online drip, micro irrigation and check basin irrigation system. The irrigation for online drip and micro irrigation was scheduled daily, based on crop evapotranspiration; and for check basin it was scheduled at 40 mm CPE with IW/CPE ratio equal to 1.2.

TABLE 9.11 Total Cost for Various Treatments (1.00 Rs. = 0.01566 US$)

Treatment	Initial cost without pump and filter unit	Fixed cost		Fixed cost of pump and filter	Total fixed cost per annum	Fixed cost per season	Variable cost		Total cost per season
		Interest on initial cost @ 10%	Depreciation on cost @ 10%				Management and input cost	Interest on mgt cost @ 10%	
						(Rs/ha)			
T1	217564.00	21756.40	19580.76	1635.40	42972.56	14324.19	57250	5725	**77299.19**
T2	99993.70	9999.37	8999.433	1635.40	20634.20	6878.07	57250	5725	**69853.07**
T3	5371.00	537.10	483.39	1278.40	2298.89	766.30	63370	6337	**70473.30**

TABLE 9.12 Benefit–Cost Analysis Of Garlic Production

Treatment	Yield				Gross return	Total cost	Net return	BC Ratio
	2008–09	2009–10	2010–11	Pooled mean	(Rs/ha)	(Rs/ha)	(Rs/ha)	
	(100 kg/ha)							
T1	46.92	28.7	61.79	45.80	229,000	77,299.19	151,700.81	**1.96**
T2	23.72	24.9	43.97	30.86	154,300	69,853.07	84,446.93	**1.21**
T3	43.96	24.69	45.46	38.04	190,200	70,473.30	119,726.70	**1.70**

Average irrigation water requirement of garlic was 87.59 ha-cm under check basin while 47.49 ha-cm under micro irrigation system. Online drip irrigation system was best for the garlic production, due to 45.78% water saving over traditional check basin method with 20 % more yield [water use efficiency of 0.95 (100 kg)/ha-cm] and having average BC ratio of 1.96.

KEYWORDS

- **BC ratio**
- **bulb yield**
- **check basin irrigation**
- **CPE**
- **garlic**
- **irrigation scheduling**
- **micro irrigation**
- **water use efficiency**

REFERENCES

1. Anonymous, (2005). *National Water Policy.* http://warmin.nic.in/writerreaddata/linkimage/nwp.
2. Doorenbos, J., & Pruitt, W. O. (1977). *Guidelines for predicting crop water requirements.* FAO Irrigation and drainage paper 24, United Nations, Rome, pp. 30–38.
3. Michael, A. M. (1974). *Irrigation: Theory and Practice.* Vikas Publication House Pvt. Ltd., New Delhi, pp. 448–581.
4. Sankar, V., Lawande, K. E., & Tripathi, P. C. (2008). Effect of micro irrigation practices on growth and yield of garlic (*Allium sativum* L. var. G.41). *Journal of Spices and Aromatic Crops*, 17(3), 230–234.
5. Shivanappan, R. K. (1998). Low cost micro-irrigation systems for all crops and all farmers. *Proc. of Workshop on Micro Irrigation and Sprinkler Irrigation System* held at CBIP, New Delhi during 28–30 April, pp. 15–20.
6. Singandhupe, R. B., Singh, S. R., Chaudhari, K., & Patil, N. G. (1998). Drip irrigation in horticultural fruit crops. *Proc. of Workshop on Micro Irrigation and Sprinkler Irrigation System* held at CBIP, New Delhi during 28–30 April, p. 54.
7. World Watch Institute (2006). http://www.worldwatch.org/food-agriculture.

APPENDICES

(Modified and reprinted with permission from: Megh R. Goyal, 2012. Appendices. Pages 317–332. In: *Management of Drip/Trickle or Micro Irrigation* edited by Megh R. Goyal. New Jersey, USA: Apple Academic Press Inc.)

APPENDIX A   CONVERSION SI AND NON-SI UNITS

To convert the Column 1 in the Column 2,	Column 1	Column 2	To convert the Column 2 in the Column 1,
	Unit	*Unit*	
Multiply by	SI	Non-SI	Multiply by

LINEAR

0.621 _____	kilometer, km (10^3m)	miles, mi _____	1.609
1.094 _____	meter, m	yard, yd _____	0.914
3.28 _____	meter, m	feet, ft _____	0.304
3.94×10^{-2} ___	millimeter, mm (10^{-3})	inch, in _____	25.4

SQUARES

2.47 _____	hectare, he	acre _____	0.405
2.47 _____	square kilometer, km^2	acre _____	4.05×10^{-3}
0.386 _____	square kilometer, km^2	square mile, mi^2 ___	2.590
2.47×10^{-4} ___	square meter, m^2	acre _____	4.05×10^{-3}
10.76 _____	square meter, m^2	square feet, ft^2 _____	9.29×10^{-2}
1.55×10^{-3} ___	mm^2	square inch, in^2 ____	645

CUBICS

9.73×10^{-3} ___	cubic meter, m^3	inch-acre _____	102.8
35.3 _____	cubic meter, m^3	cubic-feet, ft^3 _____	2.83×10^{-2}

6.10 × 10⁴ ____ cubic meter, m³   cubic inch, in³ _____ 1.64 × 10⁻⁵

2.84 × 10⁻² ___ liter, L (10⁻³ m³)   bushel, bu _____ 35.24

1.057 _____ liter, L   liquid quarts, qt ____ 0.946

3.53 × 10⁻² ___ liter, L   cubic feet, ft³ _____ 28.3

0.265 _____ liter, L   gallon _____ 3.78

33.78 _____ liter, L   fluid ounce, oz _____ 2.96 × 10⁻²

2.11 _____ liter, L   fluid dot, dt _____ 0.473

WEIGHT

2.20 × 10⁻³ ___ gram, g (10⁻³ kg)   pound, _____ 454

3.52 × 10⁻² ___ gram, g (10⁻³ kg)   ounce, oz _____ 28.4

2.205 _____ kilogram, kg   pound, lb _____ 0.454

10⁻² _____ kilogram, kg   quintal (metric), q __ 100

1.10 × 10⁻³ ___ kilogram, kg   ton (2000 lbs), ton __ 907

1.10² _____ mega gram, mg   ton (US), ton _____ 0.907

1.10² _____ metric ton, t   ton (US), ton _____ 0.907

YIELD AND RATE

0.893 _____ kilogram per hectare   pound per acre _____ 1.12

7.77 × 10⁻² ___ kilogram per cubic meter   pound per fanega ___ 12.87

1.49 × 10⁻² ___ kilogram per hectare   pound per acre, _____ 67.19 60 lb

1.59 × 10⁻² ___ kilogram per hectare   pound per acre, _____ 62.71 56 lb

1.86 × 10⁻² ___ kilogram per hectare   pound per acre, _____ 53.75 48 lb

0.107 _____ liter per hectare   galloon per acre ____ 9.35

893 _____ ton per hectare   pound per acre _____ 1.12 × 10⁻³

893 _____ mega gram per hectare   pound per acre _____ 1.12 × 10⁻³

0.446 _____ ton per hectare

ton (2000 lb) per ____ 2.24
acre

2.24 _____ meter per second

mile per hour _____ 0.447

SPECIFIC SURFACE

10 _____ square meter per
kilogram

square centimeter ___ 0.1
per gram

10^3 _____ square meter per
kilogram

square millimeter ___ 10^{-3}
per gram

PRESSURE

9.90 _____ megapascal, MPa

atmosphere _____ 0.101

10 _____ megapascal

bar _____ 0.1

1.0 _____ megagram per
cubic meter

gram per cubic _____ 1.00
centimeter

2.09×10^{-2} ___ pascal, Pa

pound per square ___ 47.9
feet

1.45×10^{-4} ___ pascal, Pa

pound per square ____ $6.90 \times^3 10$
inch

TEMPERATURE

1.00   _____ Kelvin, K
(K-273)

centigrade, °C _____ 1.00
(C+273)

(1.8 C _____ centigrade, °C

Fahrenheit,°F _____ (F-32)/
1.8 + 32)

ENERGY

9.52×10^{-4} ___ Joule J

BTU _____ 1.05×10^3

0.239 _____ Joule, J

calories, cal _____ 4.19

0.735 _____ Joule, J

feet-pound _____ 1.36

2.387×10^5 ___ Joule per square meter

calories per _____ 4.19×10^4 square centimeter

10^5 _____ Newton, N

dynes _____ 10^{-5}

WATER REQUIREMENTS

9.73×10^{-3} ___ cubic meter

inch acre _____ 102.8

9.81×10^{-3} ___ cubic meter per hour

cubic feet per _____ 101.9 second

4.40 _____ cubic meter per hour

galloon (US) _____ 0.227 per minute

8.11 _____ hectare-meter

acre-feet _____ 0.123

97.28 _____ hectare-meter

acre-inch _____ 1.03×10^{-2}

8.1×10^{-2} ____ hectare centimeter

acre-feet _____ 12.33

CONCENTRATION

1 _____ centimol per kilogram

milliequivalents ____ 1 per 100 grams

0.1 _____ gram per kilogram

percents _____ 10

1 _____ milligram per kilogram

parts per million ____ 1

NUTRIENTS FOR PLANTS

2.29 _____ P

P_2O_5 _____ 0.437

1.20 _____ K

K_2O _____ 0.830

1.39 _____ Ca

CaO _____ 0.715

1.66 _____ Mg

MgO _____ 0.602

NUTRIENT EQUIVALENTS

Column A	Column B	Conversion	Equivalent
		A to B	B to A
N	NH_3	1.216	0.822
	NO_3	4.429	0.226
	KNO_3	7.221	0.1385
	$Ca(NO_3)_2$	5.861	0.171
	NH_4NO_3	5.718	0.175
	NH_4NO_3	5.718	0.175
N	NH_3	1.216	0.822
	NO_3	4.429	0.226
	KNO_3	7.221	0.1385
	$Ca(NO_3)_2$	5.861	0.171
	$(NH_4)_2SO_4$	4.721	0.212
	NH_4NO_3	5.718	0.175
	$(NH_4)_2HPO_4$	4.718	0.212
P	P_2O_5	2.292	0.436
	PO_4	3.066	0.326
	KH_2PO_4	4.394	0.228
	$(NH_4)_2HPO_4$	4.255	0.235
	H_3PO_4	3.164	0.316
K	K_2O	1.205	0.83
	KNO_3	2.586	0.387
	KH_2PO_4	3.481	0.287
	Kcl	1.907	0.524
	K_2SO_4	2.229	0.449
Ca	CaO	1.399	0.715
	$Ca(NO_3)_2$	4.094	0.244
	$CaCl_2 \times 6H_2O$	5.467	0.183
	$CaSO_4 \times 2H_2O$	4.296	0.233
Mg	MgO	1.658	0.603
	$MgSO_4 \times 7H_2O$	1.014	0.0986
S	H_2SO_4	3.059	0.327
	$(NH_4)_2SO_4$	4.124	0.2425
	K_2SO_4	5.437	0.184
	$MgSO_4 \times 7H_2O$	7.689	0.13
	$CaSO_4 \times 2H_2O$	5.371	0.186

APPENDIX B   PIPE AND CONDUIT FLOW

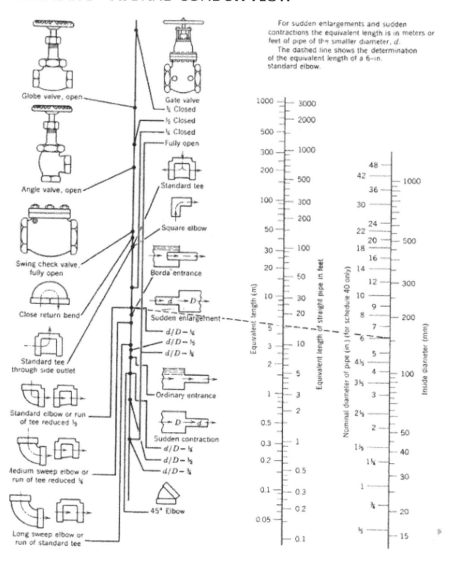

For sudden enlargements and sudden contractions the equivalent length is in meters or feet of pipe of the smaller diameter, d.
The dashed line shows the determination of the equivalent length of a 6-in. standard elbow.

APPENDIX C   PERCENTAGE OF DAILY SUNSHINE HOURS: FOR NORTH AND SOUTH HEMISPHERES

Latitude	Jan	Feb	Mar	Apr	May	Jun	Jul	Aug	Sep	Oct	Nov	Dec
NORTH												
0	8.50	7.66	8.49	8.21	8.50	8.22	8.50	8.49	8.21	8.50	8.22	8.50
5	8.32	7.57	8.47	3.29	8.65	8.41	8.67	8.60	8.23	8.42	8.07	8.30
10	8.13	7.47	8.45	8.37	8.81	8.60	8.86	8.71	8.25	8.34	7.91	8.10
15	7.94	7.36	8.43	8.44	8.98	8.80	9.05	8.83	8.28	8.20	7.75	7.88
20	7.74	7.25	8.41	8.52	9.15	9.00	9.25	8.96	8.30	8.18	7.58	7.66
25	7.53	7.14	8.39	8.61	9.33	9.23	9.45	9.09	8.32	8.09	7.40	7.52
30	7.30	7.03	8.38	8.71	9.53	9.49	9.67	9.22	8.33	7.99	7.19	7.15
32	7.20	6.97	8.37	8.76	9.62	9.59	9.77	9.27	8.34	7.95	7.11	7.05
34	7.10	6.91	8.36	8.80	9.72	9.70	9.88	9.33	8.36	7.90	7.02	6.92
36	6.99	6.85	8.35	8.85	9.82	9.82	9.99	9.40	8.37	7.85	6.92	6.79
38	6.87	6.79	8.34	8.90	9.92	9.95	10.1	9.47	3.38	7.80	6.82	6.66
40	6.76	6.72	8.33	8.95	10.0	10.1	10.2	9.54	8.39	7.75	6.72	7.52
42	6.63	6.65	8.31	9.00	10.1	10.2	10.4	9.62	8.40	7.69	6.62	6.37
44	6.49	6.58	8.30	9.06	10.3	10.4	10.5	9.70	8.41	7.63	6.49	6.21
46	6.34	6.50	8.29	9.12	10.4	10.5	10.6	9.79	8.42	7.57	6.36	6.04
48	6.17	6.41	8.27	9.18	10.5	10.7	10.8	9.89	8.44	7.51	6.23	5.86
50	5.98	6.30	8.24	9.24	10.7	10.9	11.0	10.0	8.35	7.45	6.10	5.64
52	5.77	6.19	8.21	9.29	10.9	11.1	11.2	10.1	8.49	7.39	5.93	5.43
54	5.55	6.08	8.18	9.36	11.0	11.4	11.4	10.3	8.51	7.20	5.74	5.18

APPENDIX C Continued

Latitude	Jan	Feb	Mar	Apr	May	Jun	Jul	Aug	Sep	Oct	Nov	Dec
56	5.30	5.95	8.15	9.45	11.2	11.7	11.6	10.4	8.53	7.21	5.54	4.89
58	5.01	5.81	8.12	9.55	11.5	12.0	12.0	10.6	8.55	7.10	4.31	4.56
60	4.67	5.65	8.08	9.65	11.7	12.4	12.3	10.7	8.57	6.98	5.04	4.22
SOUTH												
0	8.50	7.66	8.49	8.21	8.50	8.22	8.50	8.49	8.21	8.50	8.22	8.50
5	8.68	7.76	8.51	8.15	8.34	8.05	8.33	8.38	8.19	8.56	8.37	8.68
10	8.86	7.87	8.53	8.09	8.18	7.86	8.14	8.27	8.17	8.62	8.53	8.88
15	9.05	7.98	8.55	8.02	8.02	7.65	7.95	8.15	8.15	8.68	8.70	9.10
20	9.24	8.09	8.57	7.94	7.85	7.43	7.76	8.03	8.13	8.76	8.87	9.33
25	9.46	8.21	8.60	7.74	7.66	7.20	7.54	7.90	8.11	8.86	9.04	9.58
30	9.70	8.33	8.62	7.73	7.45	6.96	7.31	7.76	8.07	8.97	9.24	9.85
32	9.81	8.39	8.63	7.69	7.36	6.85	7.21	7.70	8.06	9.01	9.33	9.96
34	9.92	8.45	8.64	7.64	7.27	6.74	7.10	7.63	8.05	9.06	9.42	10.1
36	10.0	8.51	8.65	7.59	7.18	6.62	6.99	7.56	8.04	9.11	9.35	10.2
38	10.2	8.57	8.66	7.54	7.08	6.50	6.87	7.49	8.03	9.16	9.61	10.3
40	10.3	8.63	8.67	7.49	6.97	6.37	6.76	7.41	8.02	9.21	9.71	10.5
42	10.4	8.70	8.68	7.44	6.85	6.23	6.64	7.33	8.01	9.26	9.8	10.6
44	10.5	8.78	8.69	7.38	6.73	6.08	6.51	7.25	7.99	9.31	9.94	10.8
46	10.7	8.86	8.90	7.32	6.61	5.92	6.37	7.16	7.96	9.37	10.1	11.0

APPENDIX D PSYCHOMETRIC CONSTANT (γ) FOR DIFFERENT ALTITUDES (Z)

$$\gamma = 10^{-3} \, [(C_p.P) \div (\varepsilon.\lambda)] = (0.00163) \times [P \div \lambda]$$

γ, psychrometric constant [kPa C^{-1}]

c_p, specific heat of moist air = 1.013

[kJ $kg^{-1}\,°C^{-1}$]

P, atmospheric pressure [kPa].

ε, ratio molecular weight of water

vapor/dry air = 0.622

λ, latent heat of vaporization [MJ kg^{-1}]

= 2.45 MJ kg^{-1} at 20°C.

Z (m)	γ kPa/°C	z (m)	γ kPa/°C	z (m)	γ kPa/°C	z (m)	γ kPa/°C
0	0.067	1000	0.060	2000	0.053	3000	0.047
100	0.067	1100	0.059	2100	0.052	3100	0.046
200	0.066	1200	0.058	2200	0.052	3200	0.046
300	0.065	1300	0.058	2300	0.051	3300	0.045
400	0.064	1400	0.057	2400	0.051	3400	0.045
500	0.064	1500	0.056	2500	0.050	3500	0.044
600	0.063	1600	0.056	2600	0.049	3600	0.043
700	0.062	1700	0.055	2700	0.049	3700	0.043
800	0.061	1800	0.054	2800	0.048	3800	0.042
900	0.061	1900	0.054	2900	0.047	3900	0.042
1000	0.060	2000	0.053	3000	0.047	4000	0.041

APPENDIX E SATURATION VAPOR PRESSURE [e_s] FOR DIFFERENT TEMPERATURES (T)

Vapor pressure function = e_s = [0.6108]*exp{[17.27*T]/[T + 237.3]}							
T °C	e_s kPa	T °C	e_s kPa	T °C	e_s kPa	T °C	e_s kPa
1.0	0.657	13.0	1.498	25.0	3.168	37.0	6.275
1.5	0.681	13.5	1.547	25.5	3.263	37.5	6.448
2.0	0.706	14.0	1.599	26.0	3.361	38.0	6.625
2.5	0.731	14.5	1.651	26.5	3.462	38.5	6.806
3.0	0.758	15.0	1.705	27.0	3.565	39.0	6.991

APPENDIX E Continued

Vapor pressure function = e_s = [0.6108]*exp{[17.27*T]/[T + 237.3]}							
T °C	e_s kPa	T °C	e_s kPa	T °C	e_s kPa	T °C	e_s kPa
3.5	0.785	15.5	1.761	27.5	3.671	39.5	7.181
4.0	0.813	16.0	1.818	28.0	3.780	40.0	7.376
4.5	0.842	16.5	1.877	28.5	3.891	40.5	7.574
5.0	0.872	17.0	1.938	29.0	4.006	41.0	7.778
5.5	0.903	17.5	2.000	29.5	4.123	41.5	7.986
6.0	0.935	18.0	2.064	30.0	4.243	42.0	8.199
6.5	0.968	18.5	2.130	30.5	4.366	42.5	8.417
7.0	1.002	19.0	2.197	31.0	4.493	43.0	8.640
7.5	1.037	19.5	2.267	31.5	4.622	43.5	8.867
8.0	1.073	20.0	2.338	32.0	4.755	44.0	9.101
8.5	1.110	20.5	2.412	32.5	4.891	44.5	9.339
9.0	1.148	21.0	2.487	33.0	5.030	45.0	9.582
9.5	1.187	21.5	2.564	33.5	5.173	45.5	9.832
10.0	1.228	22.0	2.644	34.0	5.319	46.0	10.086
10.5	1.270	22.5	2.726	34.5	5.469	46.5	10.347
11.0	1.313	23.0	2.809	35.0	5.623	47.0	10.613
11.5	1.357	23.5	2.896	35.5	5.780	47.5	10.885
12.0	1.403	24.0	2.984	36.0	5.941	48.0	11.163
12.5	1.449	24.5	3.075	36.5	6.106	48.5	11.447

APPENDIX F　SLOPE OF VAPOR PRESSURE CURVE (Δ) FOR DIFFERENT TEMPERATURES (T)

$$\Delta = [4098.\ e^0(T)] \div [T + 237.3]^2$$
$$= 2504\{exp[(17.27T) \div (T + 237.2)]\} \div [T + 237.3]^2$$

T °C	Δ kPa/°C	T °C	Δ kPa/°C	T °C	Δ kPa/°C	T °C	Δ kPa/°C
1.0	0.047	13.0	0.098	25.0	0.189	37.0	0.342
1.5	0.049	13.5	0.101	25.5	0.194	37.5	0.350
2.0	0.050	14.0	0.104	26.0	0.199	38.0	0.358

APPENDIX F Continued

T °C	Δ kPa/°C	T °C	Δ kPa/°C	T °C	Δ kPa/°C	T °C	Δ kPa/°C
2.5	0.052	14.5	0.107	26.5	0.204	38.5	0.367
3.0	0.054	15.0	0.110	27.0	0.209	39.0	0.375
3.5	0.055	15.5	0.113	27.5	0.215	39.5	0.384
4.0	0.057	16.0	0.116	28.0	0.220	40.0	0.393
4.5	0.059	16.5	0.119	28.5	0.226	40.5	0.402
5.0	0.061	17.0	0.123	29.0	0.231	41.0	0.412
5.5	0.063	17.5	0.126	29.5	0.237	41.5	0.421
6.0	0.065	18.0	0.130	30.0	0.243	42.0	0.431
6.5	0.067	18.5	0.133	30.5	0.249	42.5	0.441
7.0	0.069	19.0	0.137	31.0	0.256	43.0	0.451
7.5	0.071	19.5	0.141	31.5	0.262	43.5	0.461
8.0	0.073	20.0	0.145	32.0	0.269	44.0	0.471
8.5	0.075	20.5	0.149	32.5	0.275	44.5	0.482
9.0	0.078	21.0	0.153	33.0	0.282	45.0	0.493
9.5	0.080	21.5	0.157	33.5	0.289	45.5	0.504
10.0	0.082	22.0	0.161	34.0	0.296	46.0	0.515
10.5	0.085	22.5	0.165	34.5	0.303	46.5	0.526
11.0	0.087	23.0	0.170	35.0	0.311	47.0	0.538
11.5	0.090	23.5	0.174	35.5	0.318	47.5	0.550
12.0	0.092	24.0	0.179	36.0	0.326	48.0	0.562
12.5	0.095	24.5	0.184	36.5	0.334	48.5	0.574

APPENDIX G   NUMBER OF THE DAY IN THE YEAR (JULIAN DAY)

Day	Jan	Feb	Mar	Apr	May	Jun	Jul	Aug	Sep	Oct	Nov	Dec
1	1	32	60	91	121	152	182	213	244	274	305	335
2	2	33	61	92	122	153	183	214	245	275	306	336
3	3	34	62	93	123	154	184	215	246	276	307	337
4	4	35	63	94	124	155	185	216	247	277	308	338
5	5	36	64	95	125	156	186	217	248	278	309	339
6	6	37	65	96	126	157	187	218	249	279	310	340

APPENDIX G Continued

Day	Jan	Feb	Mar	Apr	May	Jun	Jul	Aug	Sep	Oct	Nov	Dec
7	7	38	66	97	127	158	188	219	250	280	311	341
8	8	39	67	98	128	159	189	220	251	281	312	342
9	9	40	68	99	129	160	190	221	252	282	313	343
10	10	41	69	100	130	161	191	222	253	283	314	344
11	11	42	70	101	131	162	192	223	254	284	315	345
12	12	43	71	102	132	163	193	224	255	285	316	346
13	13	44	72	103	133	164	194	225	256	286	317	347
14	14	45	73	104	134	165	195	226	257	287	318	348
15	15	46	74	105	135	166	196	227	258	288	319	349
16	16	47	75	106	136	167	197	228	259	289	320	350
17	17	48	76	107	137	168	198	229	260	290	321	351
18	18	49	77	108	138	169	199	230	261	291	322	352
19	19	50	78	109	139	170	200	231	262	292	323	353
20	20	51	79	110	140	171	201	232	263	293	324	354
21	21	52	80	111	141	172	202	233	264	294	325	355
22	22	53	81	112	142	173	203	234	265	295	326	356
23	23	54	82	113	143	174	204	235	266	296	327	357
24	24	55	83	114	144	175	205	236	267	297	328	358
25	25	56	84	115	145	176	206	237	268	298	329	359
26	26	57	85	116	146	177	207	238	269	299	330	360
27	27	58	86	117	147	178	208	239	270	300	331	361
28	28	59	87	118	148	179	209	240	271	301	332	362
29	29	(60)	88	119	149	180	210	241	272	302	333	363
30	30	—	89	120	150	181	211	242	273	303	334	364
31	31	—	90	—	151	—	212	243	—	304	—	365

APPENDIX H   STEFAN-BOLTZMANN LAW AT DIFFERENT TEMPERATURES (T):

$$[\sigma*(T_K)^4] = [4.903 \times 10^{-9}], \text{ MJ K}^{-4} \text{ m}^{-2} \text{ day}^{-1}$$
where: $T_K = \{T[°C] + 273.16\}$

T	$\sigma*(T_K)^4$	T	$\sigma*(T_K)^4$	T	$\sigma*(T_K)^4$
		Units			
°C	MJ m^{-2} d^{-1}	°C	MJ m^{-2} d^{-1}	°C	MJ m^{-2} d^{-1}
1.0	27.70	17.0	34.75	33.0	43.08
1.5	27.90	17.5	34.99	33.5	43.36
2.0	28.11	18.0	35.24	34.0	43.64
2.5	28.31	18.5	35.48	34.5	43.93
3.0	28.52	19.0	35.72	35.0	44.21
3.5	28.72	19.5	35.97	35.5	44.50
4.0	28.93	20.0	36.21	36.0	44.79
4.5	29.14	20.5	36.46	36.5	45.08
5.0	29.35	21.0	36.71	37.0	45.37
5.5	29.56	21.5	36.96	37.5	45.67
6.0	29.78	22.0	37.21	38.0	45.96
6.5	29.99	22.5	37.47	38.5	46.26
7.0	30.21	23.0	37.72	39.0	46.56
7.5	30.42	23.5	37.98	39.5	46.85
8.0	30.64	24.0	38.23	40.0	47.15
8.5	30.86	24.5	38.49	40.5	47.46
9.0	31.08	25.0	38.75	41.0	47.76
9.5	31.30	25.5	39.01	41.5	48.06
10.0	31.52	26.0	39.27	42.0	48.37
10.5	31.74	26.5	39.53	42.5	48.68
11.0	31.97	27.0	39.80	43.0	48.99
11.5	32.19	27.5	40.06	43.5	49.30
12.0	32.42	28.0	40.33	44.0	49.61
12.5	32.65	28.5	40.60	44.5	49.92
13.0	32.88	29.0	40.87	45.0	50.24
13.5	33.11	29.5	41.14	45.5	50.56
14.0	33.34	30.0	41.41	46.0	50.87
14.5	33.57	30.5	41.69	46.5	51.19
15.0	33.81	31.0	41.96	47.0	51.51
15.5	34.04	31.5	42.24	47.5	51.84

APPENDIX H   Continued

T	$\sigma*(T_K)^4$	T	$\sigma*(T_K)^4$	T	$\sigma*(T_K)^4$
			Units		
°C	MJ m^{-2} d^{-1}	°C	MJ m^{-2} d^{-1}	°C	MJ m^{-2} d^{-1}
16.0	34.28	32.0	42.52	48.0	52.16
16.5	34,52	32.5	42.80	48.5	52.49

APPENDIX I   THERMODYNAMIC PROPERTIES OF AIR AND WATER

1. Latent Heat of Vaporization (λ)

$$\lambda = [2.501 - (2.361 \times 10^{-3})\, T]$$

where: λ = latent heat of vaporization [MJ kg^{-1}]; and T = air temperature [°C].

The value of the latent heat varies only slightly over normal temperature ranges. A single value may be taken (for ambient temperature = 20°C): $\lambda = 2.45$ MJ kg^{-1}.

2. Atmospheric Pressure (P)

$$P = P_o \left[\{T_{Ko} - \alpha(Z - Z_o)\} \div \{T_{Ko}\} \right]^{(g/(\alpha . R))}$$

Where: P, atmospheric pressure at elevation z [kPa]

        P_o, atmospheric pressure at sea level = 101.3 [kPa]

        z, elevation [m]

        z_o, elevation at reference level [m]

        g, gravitational acceleration = 9.807 [m s^{-2}]

        R, specific gas constant = 287 [J kg^{-1} K^{-1}]

        α, constant lapse rate for moist air = 0.0065 [K m^{-1}]

        T_{Ko}, reference temperature [K] at elevation z_o = 273.16 + T

        T, means air temperature for the time period of calculation [°C]

When assuming P_o = 101.3 [kPa] at z_o = 0, and T_{Ko} = 293 [K] for T = 20 [°C], above equation reduces to:

$$P = 101.3[(293-0.0065Z)\,(293)]^{5.26}$$

3. Atmospheric Density (ρ)

$$\rho = [1000P] \div [T_{Kv}\,R] = [3.486P] \div [T_{Kv}], \text{ and } T_{Kv} = T_K[1-0.378(e_a)/P]^{-1}$$

where: ρ, atmospheric density [kg m^{-3}]

R, specific gas constant = 287 [J kg^{-1} K^{-1}]

$T_{Kv,}$ virtual temperature [K]

$T_{K,}$ absolute temperature [K]: $T_K = 273.16 + T$ [°C]

e_a, actual vapor pressure [kPa]

T, mean daily temperature for 24-hour calculation time steps.

For average conditions (e_a in the range 1–5 kPa and P between 80–100 kPa), T_{Kv} can be substituted by: $T_{Kv} \approx 1.01\,(T + 273)$

4. Saturation Vapor Pressure function (e_s)

$$e_s = [0.6108]*\exp\{[17.27*T]/[T + 237.3]\}$$

where: e_s, saturation vapor pressure function [kPa]

T, air temperature [°C]

5. Slope Vapor Pressure Curve (Δ)

$$\Delta = [4098.\,e°(T)] \div [T + 237.3]^2$$

$$= 2504\{\exp[(17.27T) \div (T + 237.2)]\} \div [T + 237.3]^2$$

where: Δ, slope vapor pressure curve [kPa C^{-1}]

T, air temperature [°C]

$e°(T)$, saturation vapor pressure at temperature T [kPa]

In 24-hour calculations, Δ is calculated using mean daily air temperature. In hourly calculations T refers to the hourly mean, T_{hr}.

6. Psychrometric Constant (γ)

$$\gamma = 10^{-3}\,[(C_p.P) \div (\varepsilon.\lambda)] = (0.00163) \times [P \div \lambda]$$

where: γ, psychrometric constant [kPa C^{-1}]

c_p, specific heat of moist air = 1.013 [kJ kg^{-1} °C^{-1}]

P, atmospheric pressure [kPa]: equations 2 or 4

ε, ratio molecular weight of water vapor/dry air = 0.622

λ, latent heat of vaporization [MJ kg^{-1}]

7. Dew Point Temperature (T_{dew})

When data is not available, T_{dew} can be computed from e_a by:

$$T_{dew} = [\{116.91 + 237.3\,Log_e(e_a)\} \div \{16.78 - Log_e(e_a)\}]$$

where: T_{dew}, dew point temperature [°C]

e_a, actual vapor pressure [kPa]

For the case of measurements with the Assmann psychrometer, T_{dew} can be calculated from:

$$T_{dew} = (112 + 0.9\,T_{wet})[e_a \div (e° \, T_{wet})]^{0.125} - [112 - 0.1\,T_{wet}]$$

8. Short Wave Radiation on a Clear-Sky Day (R_{so})

The calculation of R_{so} is required for computing net long wave radiation and for checking calibration of pyranometers and integrity of R_{so} data. A good approximation for R_{so} for daily and hourly periods is:

$$R_{so} = (0.75 + 2 \times 10^{-5}\,z)R_a$$

where: z, station elevation [m]

R_a, extraterrestrial radiation [MJ m^{-2} d^{-1}].

Equation is valid for station elevations less than 6000 m having low air turbidity. The equation was developed by linearizing Beer's radiation extinction law as a function of station elevation and assuming that the average angle of the sun above the horizon is about 50°.

For areas of high turbidity caused by pollution or airborne dust or for regions where the sun angle is significantly less than 50° so that the path length of radiation through the atmosphere is increased, an adoption of Beer's law can be employed where P is used to represent atmospheric mass:

$$R_{so} = (R_a) \exp[(-0.0018P) \div (K_t \sin(\Phi))]$$

where: K_t, turbidity coefficient, $0 < K_t < 1.0$ where $K_t = 1.0$ for clean air and

$K_t = 1.0$ for extremely turbid, dusty or polluted air.

P, atmospheric pressure [kPa]

Φ, angle of the sun above the horizon [rad]

R_a, extraterrestrial radiation [MJ m^{-2} d^{-1}]

For hourly or shorter periods, Φ is calculated as:

$$\sin \Phi = \sin \varphi \sin \delta + \cos \varphi \cos \delta \cos \omega$$

where: φ, latitude [rad]

δ, solar declination [rad] (Eq. (24) in Chapter 3)

ω, solar time angle at midpoint of hourly or shorter period [rad]

For 24-hour periods, the mean daily sun angle, weighted according to R_a, can be approximated as:

$$\sin(\Phi_{24}) = \sin[0.85 + 0.3 \varphi \sin\{(2\pi J/365)-1.39\}-0.42 \varphi^2]$$

where: Φ_{24}, average Φ during the daylight period, weighted according to R_a [rad]

φ, latitude [rad]

J, day in the year.

The Φ_{24} variable is used to represent the average sun angle during daylight hours and has been weighted to represent integrated 24-hour transmission effects on 24-hour R_{so} by the atmosphere. Φ_{24} should be limited to > 0. In some situations, the estimation for R_{so} can be improved by modifying to consider the effects of water vapor on short wave absorption, so that: $R_{so} = (K_B + K_D) R_a$ where:

$$K_B = 0.98\exp[\{(-0.00146P) \div (K_t \sin \Phi)\}-0.091\{w/\sin \Phi\}^{0.25}]$$

where: K_B, the clearness index for direct beam radiation

K_D, the corresponding index for diffuse beam radiation

$K_D = 0.35-0.33 K_B$ for $K_B > 0.15$

$K_D = 0.18 + 0.82\ K_B$ for $K_B < 0.15$

R_a, extraterrestrial radiation [MJ m^{-2} d^{-1}]

K_t, turbidity coefficient, $0 < K_t < 1.0$ where $K_t = 1.0$ for clean air and $K_t = 1.0$ for extremely turbid, dusty or polluted air.

P, atmospheric pressure [kPa]

Φ, angle of the sun above the horizon [rad]

W, perceptible water in the atmosphere [mm] $= 0.14\ e_a\ P + 2.1$

e_a, actual vapor pressure [kPa]

P, atmospheric pressure [kPa]

APPENDIX J   PSYCHROMETRIC CHART AT SEA LEVEL

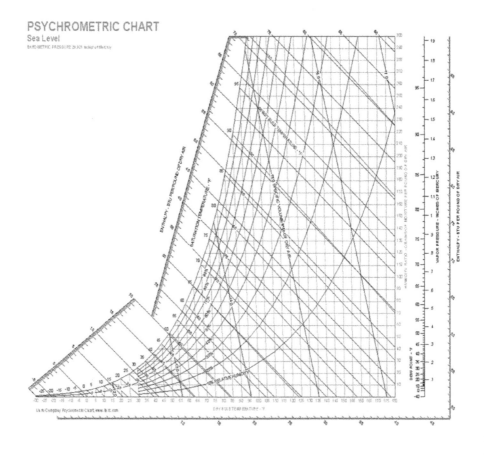

APPENDIX K

(http://www.fao.org/docrep/T0551E/t0551e07.htm#5.5%20field%20
management%20practices%20in%20wastewater%20irrigation)

1. **Relationship between applied water salinity and soil water salinity at different leaching fractions (FAO, 1985)**

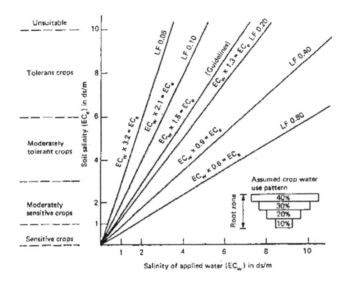

2. **Schematic representations of salt accumulation, planting positions, ridge shapes and watering patterns.**

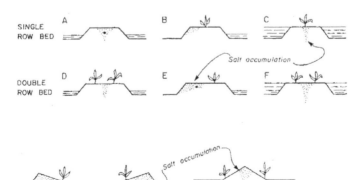

3. Main components of general planning guidelines for wastewater reuse (Cobham and Johnson, 1988)

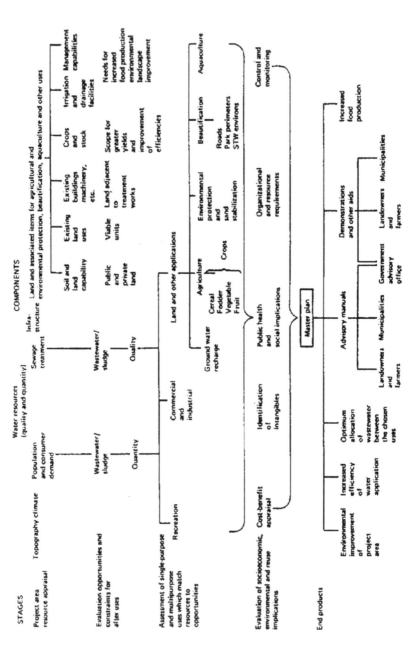

APPENDIX L

From: *Vincent F. Bralts*, 2015. Chapter 3: Evaluation of the uniformity coefficients. In: *Sustainable Micro Irrigation Management for Trees and Vines, Volume 3* by M. R. Goyal (Ed.). Apple Academic Press Inc.

1. Uniformity classification.

Classification	Statistical Uniformity	Emission Uniformity
Excellent	For U = 100–95%	100–94%
Good	For U = 90–85%	87–81%
Fair	For U = 80–75%	75–68%
Poor	For U = 70–65%	62–56%
Not Acceptable	For U < 60%	<50%

2. Acceptable intervals of uniformity in a drip irrigation system.

Type of dripper	Slope	Uniformity interval, %
Point Source: located in planting distance > 3.9 m.	Level*	90–95
	Inclined**	85–90
Point Source: located in planting distance < 3.9 m.	Level*	85–90
	Inclined**	80–90
Drippers inserted in the lines for annual row crops.	Level*	80–90
	Inclined**	75–85

* Level = Slope less that 2%. ** Inclined = Slope greater than 2%.

3. Confidence limits for field uniformity (U).

Field uniformity	18 drippers		36 drippers		72 drippers	
	Confidence limit		Confidence limit		Confidence limit	
	N_{Sum}*	%	N_{Sum}	%	N_{Sum}	%
100%	3	U ± 0.0	6	U ± 0.6%	12	U ± 0.0%
90%	3	U ± 2.9	6	U ± 2.0%	12	U ± 1.4%
80%	3	U ± 5.8	6	U ± 4.0%	12	U ± 2.8%
70%	3	U ± 9.4	6	U ± 6.5%	12	U ± 4.5%
60%	3	U ± 13.3	6	U ± 9.2%	12	U ± 6.5%

*N_{Sum} = 1/6 part of the total measured drippers. This is a number of samples that will be added to calculate T_{max} and T_{min}.

4. Nomograph for statistical uniformity

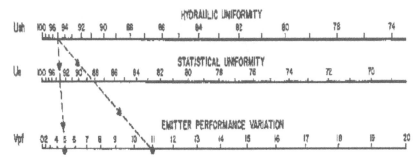

5. The field uniformity of an irrigation system based on the dripper times and the dripper flow rate.

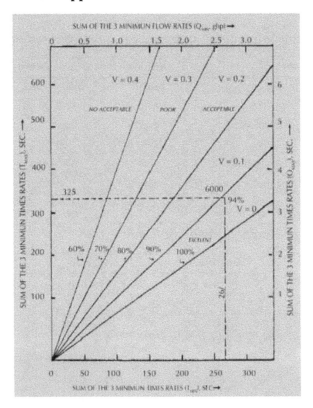

6. The field uniformity of a drip irrigation system based on the time to collect a known quantity of water or based on pressure for hydraulic uniformity.

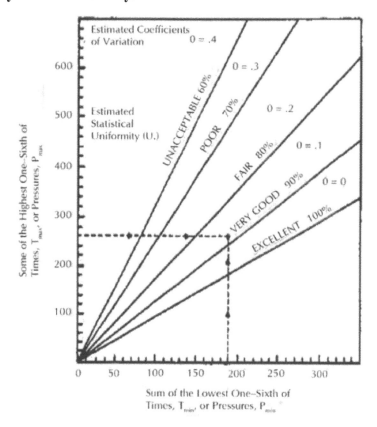

INDEX

apoplast, 89
conducting elements, 88
irrigation, 80, 101, 102
system, 89
transport, 85

Z

Zigzag path, 87
Zn/phytate ratio, 201
Zoonoses, 200
Zoonotic pathogens, 200

Milton Keynes UK
Ingram Content Group UK Ltd.
UKHW022058141024
449569UK00031B/1692